Colony: Earth

RICHARD E. MOONEY

STEIN AND DAY/*Publishers*/New York

First published in 1974

Copyright � 1974 by Richard E. Mooney

Library of Congress Catalog Card No. 73-90698

Designed by David Miller

Printed in the United States of America

Stein and Day/*Publishers*/Scarborough House, Briarcliff Manor, New York 10510

ISBN 0-8128-1658-7

To my Wife, Janet

Also,

Mother, Father, Val, David & Carole

CONTENTS

INTRODUCTION

The controversy surrounding the origins of man and civilization seems more intense today than it has ever been.

Until recently, two opposite explanations were offered. One is the religious concept of divine creation; the other is the theory of evolution through natural selection, first postulated by Charles Darwin in the nineteenth century. While the evolutionary theory probably has more adherents than the concept of divine creation, both are, for many reasons, totally unsatisfactory.

Recently, a third hypothesis has been advanced, namely, that the "gods" of antiquity were, in fact, visitors from an advanced culture elsewhere in space, and that *Homo sapiens* was created by them from existing apelike forms by the use of advanced biological engineering techniques.

But there is a fourth possible explanation: that *Homo sapiens* may not be native to this world, but *arrived here as a result of colonization*

*from elsewhere in space, and, further, that all the civilizations we know
were founded upon the wreckage of an earlier, greater culture which
was destroyed.*

Every new discovery about the past, particularly the remoter
periods, suggests a high state of knowledge at a very ancient period.

There are many examples:

• A German engineer working in Baghdad found among various
bric-à-brac in a museum articles of unknown antiquity that proved to
be parts of electric batteries.

• The Chimu of South America carried out gold plating with
results that today can be achieved only by electroplating.

• An aluminum belt was found in a grave in China. Aluminum is
extremely difficult to refine from bauxite ore, and requires a highly
complex technical plant.

• A mechanical calculating device consisting of a series of inter-
locking cogs was discovered recently in a sunken Greek galley.

• Lenses made of rock crystal have been found in the ruins of
ancient Sumer and Babylonia in the Mesopotamian deserts; examples
of microscopic craftsmanship have been found in other parts of the
world. The manufacture of magnifying instruments requires a high
degree of skill and knowledge of optics.

• In Delhi, India, there is a column of unknown antiquity, welded
together in sections and made of rustless iron. It has stood for centuries in a harsh tropical climate without a trace of rust. We are unable
to duplicate this material.

There is a legend that the pyramids contained a store of tools and weapons
made of rustless iron, and as no such relics have ever been discovered
within any of the pyramids, the legend has been dismissed. Yet ancient
rustless iron does exist; perhaps the legend is not without foundation.

Apart from actual relics, there are many stories of things which
were once regarded as myth but now have parallels in our technology
or are not beyond the capability of our science at some future date:

• There is the legend of Talos, the metal giant created by Poseidon
to guard the possessions of the Cretan kings. Was Talos a robot?

• Mythological references to "serpents' eggs" that gave forth fire and magicians' rods that spat fire, could describe explosive devices like hand grenades and rifles. Such legends are curiously akin to primitive peoples' descriptions of the "thundersticks" carried by white hunters.

Ancient knowledge of astronomy was prodigious:

• The ancient Greeks were aware that there were many more stars in the sky than could be seen with the naked eye.

• Dean Swift described the moons of Mars accurately, 150 years before the telescope had been invented to detect them.

• Babylonian priests and astronomers had charts that showed four of the larger satellites of Jupiter.

How did the Greeks know that there were many more stars than could be seen with the naked eye? From what source did Swift gain his knowledge of the moons of Mars? How could the astronomers in ancient Babylon know of the moons of Jupiter, which cannot be seen without the aid of a telescope?

Does this knowledge stem from an earlier period, from people who were in possession of the knowledge we have today? Sumerian and Babylonian astronomers talked of the myriad of heavenly *inhabited* spheres. How did they develop this idea, which we are only today beginning to take seriously?

• At Chichén Itzá in Yucatán, there is a huge building called the Observatory because it resembles nothing so much as a modern astronomical observatory. There is no equipment in this ancient building, nor is there any trace that there ever was. The roof cannot be opened, as in a modern astronomical observatory, and although there appear to be sighting holes at intervals, they do not seem to relate to any heavenly object. Nevertheless, the design of an observatory is there.

Were the ancient Maya copying something far, far older, in much the same way that we copy Roman and Greek buildings? The Maya may or may not have known what the building represented; perhaps it was regarded as a sacred structure from the vanished Golden Age. Perhaps the Maya lacked both the equipment and the knowledge for telescopic observation. Even so, they made extraordinarily accurate mathematical calculations of the moon and Venus—more accurate, in some respects, than we make today.

If we had only one of these pieces of evidence, we might be forgiven for assuming coincidence or educated guesses; but so many of them, taken together, surely suggest that at one time in the past a great body of astronomical knowledge existed, only fragments of which have survived.

These relics and legends alone point to a high degree of scientific and technical knowledge in the remote past, most of which had passed into legend by the time of the ancient Greeks, and had been forgotten before the Romans came to power.

The Greeks may have retained a measure of ancient skills: the iron column at Delhi has been attributed to the Greek incursion into India under Alexander the Great, and maps of the southern polar regions were said to have been passed down from Alexander's time. But if the secret of rustless iron was known by the early Greeks, it had been forgotten by the time the world became Greco-Roman.

Even so, were it not for the chance discoveries of a few remaining artifacts, we should dismiss the legends out of hand. In all fairness, we cannot now do so. Still, it would be difficult to imagine that these ideas were merely the work of the imagination of ancient man. *People cannot possibly imagine things of which they have absolutely no knowledge at all, and for which there is no precedent.* And in nearly all cases, the legends relate to things which we have now invented, or which are within the realm of theoretical possibility.

This book is based on facts, as far as it has been possible to verify them. The conclusions based on these facts are my own. The author is a layman, blessed (or cursed) with a great deal of curiosity and an even greater distrust of traditional, "expert" opinion.

Traditional opinion about Stonehenge, for example, *seems* simple: it was a temple of some sort built by Stone Age man to worship the sun. *But why did they bring stones such great distances, and why such huge stones? How were they transported and erected?* And if Stonehenge is such a precise mathematical device (as seems the case), *were Stone Age men the simple savages we have been led to believe?*

Could man be of extraterrestrial origin? A few years ago such an idea would have been greeted with complete scorn. But we now know that, properly equipped, man can travel in space and survive.

Astronomers are now committed to the view that planetary systems are probably the normal accompaniment of certain stars.

There is no reason to suppose that other earthlike worlds circling other stars cannot exist, and therefore that life—intelligent life—could not exist in such worlds. Of course, we have as yet no proof that life *does* exist elsewhere in the universe, but that does not mean that it does not.

Nor does the fact that we have not yet made contact with another intelligent species in the universe mean that such a species does not exist. We have had radio for less than a century; radio astronomy is even younger. It is hardly surprising that we have not yet made contact with life beyond this planet. A Superior Community transmitting over interstellar distances may well be using equipment beyond the capacity of any receivers we have, or they may be using methods of communication totally outside the scope of our present science. Perhaps some other race has already attempted to contact us, failed, and considered that we had not progressed sufficiently. It may well be that a thousand years will elapse before they try again.

Although interstellar travel is presently far beyond the capabilities of our technology, it need not be assumed that such travel is impossible. A race far in advance of our own may have long since discovered a method of traversing the vast distances involved and of circumventing the limitations of the velocity of electromagnetic radiation. Our concept of light and electromagnetic radiation as the utter limits of velocity may be in error. Some of the latest research in subatomic physics and particle states suggests that we may have to revise previously held views of the laws of the physical universe. The discovery of quasars (quasi stellar objects), for example, has confounded scientists because of their apparent disregard of all known physical laws.

The earth itself holds many mysteries for us, and man's place on earth is even more mysterious. There are many questions for which no satisfactory answers have yet been provided:

• Why did the Ice Age end? Did it end suddenly? *Or was there, in fact, no Ice Age?*

• There is evidence that vast numbers of animals died suddenly at the end of the period known as the Pleistocene. Does this point to a catastrophic end? *Do worldwide legends of a Flood in which most of humanity perished support the theory of a sudden termination of the Ice Age?* Or is there another explanation for the Flood?

• If man had evolved here from some apelike creature and then spent the following thirty thousand years living as a hunting nomad, why are legends of such a state conspicuous by their absence?

• *If man spent thirty to thirty-five thousand years as a wild hunting nomad, why did he suddenly become civilized?* Why such a long period with no progress at all, then a *sudden* flowering of culture, with cities, mathematics, plant and animal husbandry, drainage and irrigation?

• Most legends say that before the Flood lived gods, or god-men with strange powers, who could fly, and whose knowledge became so great that the gods became afraid and destroyed almost all of them in the Flood. *Who were these gods?*

• *Who were the mysterious culture bearers* who appear in the mythologies of all ancient cultures—Sumerian, Egyptian, Greek, Aryan (India), Aztec, Maya, and Inca?

• *How can we account for the worldwide distribution of megalithic structures* in remote antiquity, before the invention of the wheel? The construction of many of these monuments required high mathematical abilities and amazing engineering skills.

• *How was the city of Tiahuanaco in Bolivia built, and when?*

• *How and why was the Great Pyramid constructed?*

Today, with all our knowledge and machinery, we would be unable to duplicate this structure.

• *What is the meaning of the magician's wand, the philosopher's stone?*

• *Is the legend of Atlantis, and the myth of the Pacific land of Lemuria, a memory of civilization before the Flood?*

• *Are the legends of the Garden of Eden and the Golden Ages of many races echoes of the world as it once was, before a great catastrophe occurred?*

The concept of an older, vanished civilization exists as far back as written records exist. Much of religious tradition and literature points, even if obscurely, to such a concept.

To discuss these questions we shall have to dispense with much of the traditional approach to human affairs:

• *We shall declare that it may be that no such thing as an Ice Age occurred in the past, but that the Ice Age is now;* that the earth's orbit has altered within the past few thousand years; that some aspects of human affairs are far older than has always been thought, and that others are not nearly as old.

• *We shall say that many ancient buildings and structures were not built for the purposes which have always been thought,* and that religion did not figure nearly so prominently in the life of ancient man as has always been assumed.

• *We shall assert that what are assumed to have been the major religious texts of the world were once not religious at all,* but were a record of *actual events* which have become distorted into a religion with the passage of time.

•*We shall say that man was not always alone in the universe,* but was once associated with a *Superior Community in space,* and that this link had profound effects upon the human race.

Space travel, now in its infancy, would seem to be not so much the realization of an age-old dream (even the ancient Greeks dreamed of traveling to the moon), but almost an urge, an instinct, built into the human organism. If mankind is to exist indefinitely, then the race will eventually have to find a new home. For earth will not always be a fit habitation for man.

As the sun progresses through its hydrogen/helium cycle and becomes hotter, the earth will become more unbearable and at last unfit for human habitation. The inner planets—Mercury, Venus, earth, and Mars—will be eventually swallowed up by the expanding sun.

In the coming centuries, more pressing needs may make space travel, and particularly interstellar travel, a necessity. Overpopulation, exhaustion of natural resources and sources of power, even pollution, which may cause drastic changes to the atmosphere we breathe, may make it necessary for great numbers, if not all, of humankind to escape to other worlds.

Who is to say that such things may not have happened elsewhere in the galaxy in times past, forcing other races to flee their homes, to settle other planets? Perhaps survivors of some menacing situation on another world found it necessary to colonize earth.

ONE OF MILLIONS

We have been taught that man is very special, created by God in his image to rule the world. As man was regarded as the apex of God's creation, so the world was regarded as the center of the universe, the planet specially created as man's home. The sun, the moon, and the stars revolved around it.

With the development of the astronomical sciences, it was found, and had to be admitted by both church and secular authorities, that earth did not occupy a special place at the center of creation, and in fact was not at the center of anything at all. In fact, earth is, we now know, the third planet in distance from an ordinary G-type star, indistinguishable from many thousands of its kind.

The position of our star has been established by radio astronomy, which has also given us a guide to the structure of the galaxy of which our sun is a member star. Viewed from outside, the galaxy looks like a huge pinwheel of stars, with a nucleus surrounded by spiral arms of dust, gas, and star clouds.

The galaxy is thought to be approximately 100,000 light-years in diameter, and 10,000 light-years thick toward the nucleus. Light, traveling at 186,000 miles per second, traverses roughly 6 trillion (million million) miles in a year; astronomical distances are generally reckoned in light-years, or parsecs (3.26 light-years). Therefore, when we talk of light-years we mean multiples of 6 trillion miles, and of parsecs, multiples of some 19.2 trillion miles.

The galaxy rotates slowly about its axis, the nucleus, once in 220 million years, a period known as a cosmic year, and is thought to be between 10 and 25 billion years old. It is estimated to contain between 50 and 150 billion stars, in addition to vast clouds of dust and gas. Satellite galaxies, the globular cluster in Hercules and the irregular star clusters called the Lesser and Greater Magellanic Clouds, accompany the galaxy in much the same way that planets accompany the sun.

The sun and its retinue of planets, moons, asteroids, and meteoric dust are situated in one of the spiral arms (of which three so far have been detected by radio astronomy), some 27,000 light-years from the galactic nucleus. The more accurate location of the sun is known to be the inner edge of the Carina-Cygnus arm, on a spur that juts toward a group of hot stars in Orion. This aggregate of stars forms a bun-shaped knot 1,000 parsecs across, known as the Local Cluster. Our sun is some 100 parsecs from the center of this cluster.

This galaxy, whose size is staggering to the imagination—the figures are so great as to be virtually meaningless—is only one out of many millions of similar galaxies. Indeed it has been reckoned by some authorities that there are as many galaxies as there are stars in our galaxy: 100 billion.

These discoveries were a great blow to human pride, and as one misconception was shattered, another was erected to take its place and restore the idea of the unique, special creation of the home of man. Earth, said the wise man, is not, after all, the center of the universe. But it is still unique.

Thus was born the theory of the formation of the solar system by a cosmic collision. The best-known version of this theory was postulated by Sir James Jeans and H. Jeffreys in 1917. In brief, this theory holds that a passing star raised a tide on the surface of the sun and pulled away a long filament of gas. This gas, still held by the gravitational attraction of the sun, condensed into blobs and took up orbits, forming the planets and satellites we are familiar with.

The "cosmic collision" theory has serious flaws, the main one being that such a gaseous filament is likely to dissipate in space before it has time to condense into globules. Furthermore, the angular momentum would not have been great enough to form planets as distant as Pluto.

Any hypothesis based on colliding stars is not very tenable, particularly in the thinly populated region occupied by our sun. The mass of the galaxy rotates, taking with it its clouds of stars, and it is extremely unlikely that stars are going to be wandering about haphazardly to collide with other stars. They follow their orbits about the galactic nucleus in much the same way that planets orbit the sun. It is almost like saying that Mars may suddenly turn from its orbit and collide with earth.

According to the collision theory, we would eventually be left with a galaxy in which only a few thousand—if that many—stars would be attended by planetary systems. By a further process of elimination, it would be virtually impossible that among such planetary systems an earthlike, life-bearing planet would occur. Ultimately, then, this thesis satisfies the problem of the uniqueness of the earth.

Modern astrophysics leads us to a completely opposite point of view. But before we proceed further we must delineate the type of stars most likely to be attended by planetary systems.

Stars are catalogued by letter, followed by a number (except for the very hottest stars, which are known as Wolf-Hayet stars), then, in descending order of heat: B the hottest, A slightly less so, then F, G, K, M, R, N, S. These letters denote the spectral type. Numerical classification runs from 0 to 9. For instance, the sun, as a G0, is hotter than Tau Ceti, which is G4, and lower in temperature is Sigma Draconis, which falls lower than Tau Ceti at G9.

We are concerned with what are called solar-type stars, which fall into the classifications G, K, and M, categorized as main-sequence stars. The energy output of stars in the main sequence is stable over periods of hundreds of millions of years. Of course, great fluctuations in temperature would mean that life would be unlikely to develop or survive on planets of such stars. It is also unlikely that fast-spinning stars with high surface temperatures would develop planetary systems.

To understand why solar-type stars on the main sequence are most likely to be attended by planetary systems, we must review the mechanism and evolution of stellar bodies. Stars are formed from aggregates of interstellar dust and gas, compressed by gravitational attraction. When the compression advances beyond a certain point, a thermonuclear reaction is initiated, and the dust-gas agglomeration, which until now has been dark, begins to shine with its own light. Eventually the whole mass glows brightly and the protostar is born.

Using our own sun as a model, we may reconstruct its early history according to modern astrophysical research. In its first stages, shortly after the aggregate coalesces (reckoned in the case of our sun to be some 5 billion years ago), the protosun shines brightly through the halo of gas and dust surrounding it. It collects this gas and dust as it passes through space on its slow orbit around the galactic nucleus in company with other stars.

Gravitational compression maintains the thermonuclear reaction, converting the basic fabric of all stellar material (hydrogen) into helium, and the spin (centrifugal force), which is always developed within gravitational fields, causes an ejection of a quantity of gases from the star. These gases, together with the gas and dust collected by the star's gravitational field, form a spiral halo around the center of gravity of the star.

In the formative stage, then, the solar system resembles a miniature version of the galaxy, a flattened spiral of dust and gas surrounding a central nucleus. Unlike the star, this halo of dust and gas cannot initiate intense thermonuclear reactions. It therefore remains much cooler, and compresses into globules which become the protoplanets and moons.

Rapidly spinning stars are more likely to expel the surrounding gas and dust at great velocities so that they will be dissipated in space. Therefore we do not expect planetary systems to form around such stars. Stars with a low axial velocity in the spectral classes G, K, and M are, then, most likely to be attended by planetary systems.

This hypothesis of planetary formation, known as the cold agglomeration theory, seems, in the case of our own system at least, to make the most sense. The planets all lie in the same approximate plane of the elliptic, they rotate in the same direction as the sun (with the exception of Venus, Uranus, and the outer satellites of Jupiter), and their sizes and distribution fit with an origin in a spiral disc of gas and dust.

So when we search for planetary systems, we are seeking stars that fall into the spectral classes G, K, and M and have slow axial rotation.

Can we detect such stars? We can certainly classify a great many of the correct spectral class, and they seem to be extremely common, even in the sparsely occupied region of our sun. The difficulty is in determining whether they can be attended by planetary systems.

When we realize that the mass of the entire solar system of planets, moons, and asteroids is only *2 per cent* of the mass of the sun and that the sun's volume is more than a million times larger than that of the earth, then the difficulty of locating planets by optical means when the star itself shows only as a point of light may be readily understood. The fact that planets have no light of their own, and shine only by light reflected from the primary—the sun of any planetary system—adds to the difficulty. Optically, it would be impossible to detect an earth-size planet orbiting even the nearest star, Alpha Centauri.

There is one method by which this difficulty may be resolved, although it requires extremely delicate measuring techniques to provide an element of proof, and even so can be applied with any measure of success only to stars that are—as astronomical distances go—fairly near.

Careful examination of the sun has shown that its axial rotation is not completely stable; that is, the sun displays a slight "wobble" some million miles from its center, a point that coincides not with the center of the sun but with the center of the solar system as a whole. This means that the planetary system, small though its mass is in comparison with the primary, *does* exert a certain degree of influence. Naturally, compared with the size of the sun, the movement is extremely small, but it has led astronomers to seek such a movement among the nearer suitable stars.

Minute measurements over a period of time have revealed that such movements seem to occur among the cooler and slower-rotating stars (G, K, M) in our vicinity. The existence of such movements does not prove the presence of planetary systems—it could denote the possibility of a dark companion star—but the fact that the same perturbations that affect the sun apply also to other, similar, stars increases the probability that planetary systems other than ours do exist.

The following stars exhibit the perturbations described and therefore indicate possible planetary systems:

Star	Spectral Type	Distance in Light-years	Surface Temp.
Sun	G0	—	5,700 k.
Barnards Star	M5	6.	2,000 k.
Lalande 21185	M2	8.2	2,870 k.
61 Cygni	K5, K8 (binary)	11.1	4,200 k.
			3,400 k.

Planetary systems have not been identified by this method for the two stars in our neighborhood that would be most likely to have such systems, Alpha Centauri, a G0 sun, 4.3 light-years distant, and Tau Ceti, almost the same surface temperature as the sun, a G4 star, at a distance of 11.8 light-years.

A hypothetical planet similar to earth would, of course, have to be situated nearer to, or farther from, the primary, depending on the surface temperature. For example, an earth-type world could well exist somewhere between 90 and 140 million miles from Alpha Centauri, but a similar world circling Barnards Star would do so at a distance of some 30 million miles.

The sun is situated in a thinly populated sector of the galaxy—the star density is approximately one for every ten light-years against one per light-year (a factor of ten) in the more densely populated regions such as the Sagittarius Cloud. The possibility of other planetary systems would seem to be quite high, particularly in these more densely populated areas, and the greatly reduced distances would make detection of intelligent life forms on such systems far easier.

2

WAS THERE LIFE
BEFORE EARTH?

All forms of terrestrial life are composed of elements common throughout the universe. In fact, the commonest elements, hydrogen and oxygen, which comprise most life forms on earth, are also two of the commonest elements in the composition of both stellar bodies and the gas and dust of interstellar space.

The distinction between life and nonlife is very slight. Viruses, for example, neither grow nor move when inactive, in which state they resemble inert chemicals. Chemically, viruses have been reconstituted in laboratory experiments. When active, attacking a host cell, they react chemically, like crystalline formations. They do not reproduce by mitosis (asexually) to increase their numbers, but convert the contents of the cell by chemical means to create more viruses.

However, the action of a virus, in attacking a cell to replicate its species, is more purposive than a mere chemical reaction, and it is in this respect that the virus resembles a form of life. Furthermore, by its high degree of host specificity—viruses are extremely differentiated as to which organism they will invade—the virus would seem to be a late development form, and not a primitive forerunner of life.

There are two principal factors which differentiate life from any other system in nature, chemical or nuclear. All systems, apart from life, *devolve from the complex to the simple:* complex chemicals tend to break down into their constituent parts in time, rocks wear away, are ground into sand, to be compressed and worn away again.

Life systems operate in the reverse manner. They start from simple cell structures, which divide by mitosis, and the system keeps growing until the complete specific organism assumes an independent existence. It continues to grow until maturity, then gradually ages until death overtakes it. Once again it becomes an inert chemical system and is gradually broken down into its constituent elements. While it is alive, however, it defies what is called the "arrow of time," which leads one way toward simplification, entropy, and dissolution.

Secondly, there seems to be more to life systems than that they grow, absorb nutrients and eliminate wastes, reproduce, and move of their own volition. What distinguishes life from all other systems seems to lie in its *purposive behavior*.

Recent discoveries in the field of life studies, the now famous "coil of life," DNA (deoxyribonucleic acid), with its complementary "messenger system," RNA, have shown that all forms of life grow according to a genetic code implanted in the DNA helix. The DNA/RNA complex acts rather in the same way as the magnetic tape in a computer. As the double coil unwinds, it replicates cell construction in a step-by-step method which eventually builds the complete organism. The creature is a perfect replica of its parents and inherits characteristics from both, which are incorporated in the genes donated by the father and the mother.

The discovery of DNA, with all its implications, was seized upon by believers as further proof of the existence of God. In the case of the computer, we know that someone has programed the information onto the tape for it to act upon. Could it not be said, then, that as the computer has a programer, the DNA helix also has a planner? Perhaps the concept of a "vital force" which "animates" chemical systems, making them life systems, could be expressed in a different way.

The advent of nuclear physics has brought to light a host of atomic particles whose existence was formerly unsuspected by physicists. Earlier, it was thought that there were protons (positively charged particles), electrons (negatively charged), and neutrons (no charge). This view of the atomic system, rather like a solar system in miniature, has now changed. We have now, it appears, many particles, including the mysterious *neutrino,* which has no charge and a mass only one-tenth that of the electron, and can pass through solid matter as if it did not exist. The neutrino cannot be "seen," but its presence can be inferred from its interaction with other particles.

Sir Arthur Eddington theorized in the nineteen-thirties that there could be another particle of a similar order to the neutrino, which could be called a *mental particle,* or *"mindon."* Lawdon, of the University of Canterbury, New Zealand, proposed that mind is a universal property of matter, and V. A. Firsoff theorized that mental particles interact to form "mental entities" which govern or control purposive action in organic chemical systems.

Such a proposition means that the brain is the seat of the *mind,* which is actually a separate organ *operating through the brain.* We might make the somewhat weak analogy that a highly complex machine can make what may seem almost intelligent actions as long as electricity is put through its circuits. The electricity is part of the machine while it is functioning, yet it is still a separate energy flow in its own right.

Intensive research carried out in recent years into the parapsychological field of mental states seems to support the view that mental interactions, or mind particles, would operate on a different level from ordinary particles, in the same way that neutrinos disobey the laws of physics. Perhaps they operate in a different space/time continuum, and therefore cannot be detected by ordinary means.

What we are attempting to do is to give credence to the concept that planetary systems are a normal complement of solar-type stars, with the further implication that as such systems are formed by universal rules, so life will follow the same pattern of biophysical rules whether it be on earth or elsewhere in the universe.

If we are led to the assumption that all solar-type stars, including our own sun, are made from the same materials by the same processes and evolve planetary systems in the same manner, then it must also follow that the life which has arisen here on earth must also have arisen elsewhere under similar circumstances. It cannot be said that the processes which would create the one would not create the other.

The planet earth has been estimated to have existed for some five and a half eons (an eon is a billion years).

Life on earth, as far as we have been able to ascertain, has existed for some two and a half eons. So for at least half of its existence, the earth was a completely dead world. There is a certain mystery connected with the earth's earliest phase. A considerable body of evidence seems to show that for the first two eons of its existence, it had no axial rotation and held one side always to the sun, as the moon does to earth. What forces were involved in developing axial spin we do not know.

The recent moon landings and examination of lunar rocks have led scientists to theorize that the moon was not a part of earth in the remote past, as had been thought, but is perhaps even older than earth. The moon may be a captured body; perhaps this capture explains the axial spin of earth.

Although we can hazard a guess as to when life gained a foothold on the planet, we do not yet know how or why. Several hypotheses have been advanced in explanation, and although we shall be examining these at greater length toward the close of this book, we will touch briefly on several of them here.

The most popular at present is the *probiotic soup concept*. During earth's earliest years, warm oceans were rich in minerals, and huge electrical discharges were common. In laboratory experiments, a reconstruction of the supposed chemical composition of these early seas, when electrical currents are passed through them, produces amino acids, the basic "building blocks" of life systems. It is thought that huge electrical discharges—lightning perhaps—passed through the chemicals of the early seas would have had the same effect. Over millions of years, permutations and endless combinations of amino acids, protein chains, sugars, and salts would eventually produce the first unicellular forms of life.

According to the *panspermia hypothesis*, dormant spores or seeds are carried upward through planetary atmospheres and escape into space, whence they are carried by light pressure to the atmospheres of other worlds, and germinate on their surfaces. This theory, although first suggested in a more primitive form by the Greek philosopher Anaxagoras (500–428 B.C.), owes much of its present form to Svante Arrhenius (1859–1927). The main difficulty with this theory is that the direction would probably be outward, away from the sun. This means that life would have arisen first on either Venus or Mercury, which, as we shall see, is highly improbable. Also, it is difficult to see any form of life of our type being able to survive the high temperatures and radiation so near the primary.

A variant on this theory is that life may have been brought to earth by *meteoric objects*. Although the great heat developed by the passage of a meteorite through the atmosphere might be fatal to any spores it carried, it is possible that the spores or seeds may have separated in the upper reaches of the atmosphere before the frictional heat became too great, and drifted down to the surface. The discovery of traces of cellular structures, "organized elements," in certain meteorites (there is a great deal of controversy over this), lends a degree of weight to this theory.

The origin of meteoric and asteroidal material is a subject for much debate. Some scientists hold that meteoric debris and the asteroids may be the remnants of a planet that disintegrated some time in the past. Curiously enough, according to Bode's law of planetary distances, there should be a planet between the orbits of Mars and Jupiter, a space now occupied by the asteroid belt.

Either there was a planet that disintegrated or the debris in this locality failed to form an agglomeration of planetary dimensions. It seems curious that this would have happened in just this one instance. Perhaps the close proximity of Jupiter and its strong gravitational field prevented the formation of the planet. However, as Jupiter possesses a large number of satellites, some of which are quite large, this seems unlikely.

The composition of meteorites has failed, so far, to tell us whether they originated as a result of a disintegrated planet or not. Radioactive dating techniques place the age of meteorites at about four and a half eons, roughly the same as earth's. Stony and iron meteorites point to an origin in a planetary body of some size. The reported discovery of diamonds in some meteorites indicates that these must have been formed under great pressure, which also suggests a planetary origin.

On the other hand, the structure and texture of chondrite meteorites suggests that they were not formed within a large gravitational field. The chondrite meteorite which landed in Orgueil in France in 1864 has provided the biggest puzzle of all. In 1961 Claus and Nagy discovered the "organized elements" in this meteorite: microscopic spheres, spiny forms, shield-shaped bodies, a hexagonal body with three tubular protrusions, and some cylindrical forms. Ross discovered collapsed spore membranes, and two bodies of a mushroom shape.

When it was claimed that these were fossil organisms, the idea was laughed at in more dogmatic circles, who claimed that the meteorite had been contaminated by terrestrial forms after it landed. Staplin, however, traced within the hexagonal bodies minute particles of black pyrito-organic material, a mineral common in fossils but not found in recent biological contaminations.

Meteorites may in fact contain biological materials in a fossilized state. Furthermore, these meteorites may have formed part of a planet lying between the orbits of Mars and Jupiter which was shattered an estimated 500 million years ago. The distance of such a hypothetical planet from the sun, particularly if it had a dense atmosphere to retain heat, would not rule out a pattern of life like that on earth.

Could such a planet have held intelligent life? If so, it is possible that this intelligent life, either accidentally or deliberately, seeded earth with life through space probes. Perhaps such an intelligent form had commenced exploration of nearby worlds, as we are now doing with the moon, Venus, and Mars, and accidentally "contaminated" earth, as we may have done when a Russian Venus capsule crash-landed on that planet's surface.

Or perhaps the "contamination" was deliberate. If they knew their world was going to be destroyed, is it not possible that they seeded another world with life before their own world's end? Perhaps they chose Venus and Mars as well as earth. On earth the seeds found fertile ground and flourished. What about Venus and Mars? We do not know.

Impossible? We cannot be too sure. We are doing these things now, and what we are doing, others may have done in the far distant past. The possibility also exists that they seeded deliberately to make earth a suitable world for later colonization. We are considering seeding the clouds of the upper atmosphere of Venus with micro-organisms capable of converting the atmosphere from a carbon dioxide to an oxygenized atmosphere, which would eventually transform the entire nature of the planet into something more suitable for human habitation.

Was earth, perhaps, "brought to life" through such planetary engineering by intelligent beings who were not supernatural but tangible and real? We cannot be sure it was not so.

On the other hand, our hypothetical planet may have met its end in a cosmic catastrophe; some of its fragments may have landed here accidentally, to blossom to life on their new world. Of course, for such a theory to be tenable, either life started here much later than we had thought, or earth was seeded from this hypothetical planet much earlier than its estimated destruction 500 million years ago.

The idea of life seeded from elsewhere in space has several further implications: it may have drifted, or been created, in the void of interstellar space, filtering down through the atmospheres of suitable worlds. Or it may have reached earth in probes launched by exploring races in other parts of the galaxy, perhaps from even farther away.

Impossible? Under serious consideration as a future project by Wernher von Braun is launching an automated probe in the direction of the Centaurus system. Such a probe, although it may take many human lifetimes to reach its destination, may carry small organisms like viruses which could survive in an inert state for such a period of time. If an earth-type world exists in the Centaurus system, such micro-organisms may well spring to life.

It is true that none of these ideas of extraterrestrial origins for life explain *how* life originated in the first place. But if one of these ideas could be proved to be true, it would at least solve the riddle of the origin of life on earth, which, after all, is our main interest. The same arguments will apply to man himself, as we shall see.

Also, any of these ideas, with the exception of the probiotic soup theory, would explain the sudden appearance of life on earth. (Incidentally, the concept of divine creation would also explain the sudden appearance of life.)

Regardless of how life did arise on the planet, the picture which emerges follows roughly this pattern:

The earliest traces of life found to date are blue-green algae found in rocks two and a half eons old. Algal cells, fungoids, and what appear to be bacterial cells have been found in rocks two eons old; the first soft-bodied animals, two of which bear no resemblance to any fossil or living forms, have been identified in the lower Paleozoic layers (one to one and a half eons ago). In recent years, then, we have pushed back the frontier of life from the Paleozoic, one and a half eons ago, to two and a half eons ago. If further discoveries were made that doubled the time factor, it would be difficult to decide which came first—life, or earth itself!

After its appearance, life underwent increasingly complex transformations. It is thought that the first stage of complexification was the banding together of separate cellular organisms, perhaps similar to primitive amoebas, into colonies, where, in the course of time, groups of cells became differentiated into areas of specialized function. (It is interesting in this connection that cell tissues from present living forms can be kept alive and healthy in suitable nutrients, which seems to point to a period when life existed in a unicellular state when necessary, and that this faculty still remains.)

There is a great—and puzzling—gap between these early traces and the next stage, the Cambrian period, some 600 million years ago. At this time there were the trilobites and ammonites, hard-shelled crustacea, together with a variety of smaller creatures and plants, living in the warm seas of a much hotter earth. The first plants had begun to conquer the land surfaces, until now bare and lifeless. At this stage, apparently, the main groups of phyla had already been differentiated, but it has not been possible to trace the steps that led to this.

In the epochs that followed, first came the fishes, and then the primitive amphibia. At this time, animal life still lived only in the seas, and not until plant life gained a firm hold on the dry land surfaces of the earth did the animals begin to leave their watery environment. Plant life emerged into the open air on the land prior to the emergence of animal forms. This leads to a very interesting and significant point.

As a carbon-based life form, animals were dependent on oxygen to survive. At a very early period in earth's history, all animal life lived under the water, possibly because it was the only place oxygen was available. The primitive atmosphere of earth may have been similar to that of Jupiter and Saturn. Until plants gained a hold on the land there would be no oxygen in the atmosphere. Furthermore, the absence of free oxygen at this stage, far from assisting life to develop, would have hindered the early stages.

The mechanism of plants, which absorb carbon dioxide and release oxygen by photosynthesis, would alter earth's primitive atmosphere from a reducing one to an oxidizing one. It has been estimated that it took one eon for the atmosphere to be converted from its main constituents, ammonia and methane, to its present form, through the action of green plants. Not until this process was accomplished could animal life leave the water and live on the land surfaces.

Animal life did not appear on the surface areas of earth because it was a suitable place for life; rather, life, once it gained a foothold, began the process of adapting the planetary conditions to suit its needs.

Once life gained a hold on the land, it underwent many variations and complications and, in addition to the reptilian forms, developed many species of insects. The long period of the Mesozoic (middle life) Age, saw the reign of the dinosaurs, the huge reptilian orders. They died out, and were replaced by the mammalian forms of the Cenozoic (recent life) period, some of which flourished during the period dominated by the reptiles.

Why the reptilian orders came to an end after so many millions of years is still an unresolved mystery. Changing climates, with the first appearance of desert areas and perhaps much colder conditions, may have been a major factor. Reptiles may not have been able to adjust to the lower temperatures, although we are unable to state categorically that they were cold-blooded, as are modern reptiles. As we have only their skeletons, there is no way of telling if this was actually so.

An alternative suggestion is that a supernova at not too great a distance from our sun may have sprayed earth with deep-penetrating hard radiation and killed off the reptiles. But why would only the reptiles have been affected? Mammalian forms survived and eventually took over earth, and many species of insects have survived from the Paleozoic to the present day, some almost unchanged.

What caused these widespread changes of life forms on earth?

All we can be sure of is that life is here, on earth, with three major possible theories for its origin:

1. It arose spontaneously from chemical reactions.

2. It arrived here in the form of seeds or spores from elsewhere in space.

3. It was created miraculously by a divine agent.

3

IS THERE LIFE ELSEWHERE?

If life adapted earth to its needs, and not the other way around, could this also have happened on other planets in the solar system? We do not yet know.

Man's visits to the moon have been short and limited to a few square yards of the lunar surface. Lunar soil samples reveal no trace of organic material (though they do have the curious property of being able to kill terrestrial bacteria). However, this does not necessarily mean that the moon is completely dead. We do not know what conditions may be at the bottom of the deeper craters, among the mountains, or in fissures or caverns under the surface.

On the face of it, the moon seems an unlikely place for life to exist. Apart from the total lack of atmosphere, the vast changes in temperature between day and night would make any form of life as we know it unlikely. Even so, there are primitive forms of life here on earth that can survive without air and water. Some terrestrial bacteria can adapt to a surprising range of temperatures—some, for example, live in the boiling waters of hot springs, and others have adapted to the subzero temperatures of Antarctica.

We can be fairly sure that Mercury, only 36 million miles from the sun, is a dead world. The sunward side has temperatures around 700°F., and the dark side has temperatures close to absolute zero. There is a twilight zone between the dark and light faces, but whether any form of life could have developed and survived in such a limited environment is highly dubious. The outer planets—Saturn, Jupiter, Uranus, and Neptune, known as the Jovian worlds—could not support life of a terrestrial pattern.

Jupiter, the largest planet in the system, has a mean diameter of 88,700 miles; its volume is 1,312 times as large as earth. The atmosphere consists mainly of ammonia and methane, and the depth of the atmosphere, some 10,000 miles, produces pressure at surface level on the order of tens of thousands of tons per square inch.

In spite of its enormous size, the mean density of Jupiter is light in comparison to its bulk, and it is thought that the planet actually consists of a small rocky core surrounded by layers of solid hydrogen, with seas of liquid ammonia and methane. It has been thought that the "surface" is actually a continuation of the atmosphere, rendered more liquid or solid as the pressures increase toward the center of gravity. If any form of life has arisen on Jupiter, it would have to follow a vastly different chemical constitution than that of terrestrial forms, and either be able to withstand enormous pressures and gravities or have developed high in the planetary atmosphere.

Jupiter has twelve satellites, two of which, Ganymede and Callisto, are larger than either Mercury or Pluto. While there can be no certainty of an atmosphere on Callisto, Ganymede has an atmosphere and two distinct polar caps, although the composition of its atmosphere has not been verified. If life of any type we can visualize exists in the Jovian system, it is far more likely to have developed on Ganymede than on Jupiter itself.

Saturn is very similar in its composition and atmosphere to Jupiter. Here again, the atmosphere is mainly ammonia and methane, with enormous pressures at the bottom of thousands of miles of thick atmosphere. Saturn, of course, is famous for its ring system, which may be composed either of ice or of the debris of a shattered satellite.

Saturn has nine moons, the largest of which is Titan, slightly larger than Mercury. Titan's atmosphere is similar in composition to Saturn's.

The atmospheric composition of Neptune and Uranus, situated at vast distances from the sun, is thought to be similar to both Jupiter and Saturn, although in the case of Uranus there is little ammonia, and a great deal of hydrogen and helium. The surfaces of these planets may be covered in frozen or liquid methane.

Very little is known about Pluto. It may be as large as earth or a little smaller, and it lies so far from the sun that it would show only as a bright star does from earth. Pluto is permanently frozen, and could not support life as we know it, and probably as we do not know it.

We are left with only two real possibilities: Mars and Venus. If one applies the same rules of planetary development we use for earth, it is difficult to understand why at least fairly similar conditions do not apply to Mars and Venus. Like earth, they are in the "zone of life" for a star the size and temperature of our sun. Since Mars is smaller and has a lower gravity than earth, it has been theorized that there would be a greater tendency for its atmosphere to escape into space. Because it is smaller, it would also have cooled much earlier, with consequent earlier development and earlier "death." Mars may now be what earth may become in old age.

Recent photographs of Mars taken by Mariner spacecraft have shown a surface more like that of the moon than that of earth. It now appears that the famous "canals" are an illusion created by chains of craters very similar to those of the moon. Unlike the moon, however, Mars does have an atmosphere, although it is far more tenuous than previously supposed. Mars also possesses white polar caps, possibly of hoar frost and of no great depth; they shrink rapidly with the onset of summer in the appropriate hemisphere.

However, strong evidence from absorption bands in the spectrum suggests that organic compounds of some sort do exist on Mars; it is possible that there are lowly forms of life on that planet. The size and orbital positions of Deimos and Phobus, the planet's two tiny satellites, are so peculiar that it has even been thought they may be of artificial origin. Probably the problem of life on Mars and the riddle of its moons will be solved only when a manned expedition explores the Martian system.

Venus presents an even greater mystery. Because of its thick cloud covering, no details of the surface can be seen. Automatic probes that have approached Venus closely have sent back confusing data. They indicate that the atmosphere of Venus is composed mainly of carbon dioxide, ruling out the possibility of photosynthesizing vegetation on the Venusian surface. The surface temperature has been estimated at about 800 degrees F. With these temperatures, a carbon dioxide atmosphere, and no water vapor, Venus appears to be a swelteringly hot, waterless desert eternally blown by hot gales. A Russian space capsule, successfully soft-landed on the surface, tended to corroborate these findings, but ceased transmission almost as soon as it landed.

Data provided by high-altitude balloons gives a completely different picture. An unmanned space observatory at a height of 86,000 feet above earth confirmed the presence of water vapor in the form of ice crystals in the upper layers of the Venusian atmosphere, and W. Sinton has found the temperature of the upper layers of the cloud cover to be —40 degrees C.

Four models have been suggested:

1. The hot, dry, windblown desert, sweltering under an 800-degree Fahrenheit heat, with skies perpetually clouded with a thick mantle of carbon dioxide.

2. A planet covered completely in seas, and swarming with Paleozoic-type life forms.

3. The Carboniferous jungle environment.

4. A planet covered in petroleum oceans composed of condensed hydrocarbons.

It will be some time before we solve the problem of life on the inner planets of this system and why such great differences exist between them.

Interestingly, earth seen from a distance of 200,000 miles appears almost as heavily covered in clouds as Venus; furthermore, *out of a quarter of a million photographs taken by the Tiros satellite only one showed any evidence of intelligent life on the surface of our world.* These observations were made from distances much closer than any prolonged studies we have been able to make of any body in the solar system except the moon.

Unless Venus does bear life forms similar to those of earth, it seems that a hominid species may not have existed in this solar system except on earth.

If Homo sapiens *had an extraterrestrial point of origin, it is most likely the home world was a satellite of another solar-type star.*

4 DID WE COME FROM THE STARS?

"What humanity was like several thousand years ago and what it will be like in a few million years—all this, according to the theory of probability, can be found in the planetary world."

—KONSTANTIN E. TSIOLKOVSKY (1857–1935),
"The Father of Russian Rocketry"

"Perfectly ridiculous, unspeakably anthropocentric, hopelessly vain are those who believe that we are something special and superior in a universe of 100,000 million billion stars."

—Professor HARLOW SHAPLEY
Harvard College Observatory

We have seen that it now appears planetary systems are the rule, that the suns and their planetary systems are formed of the same materials as our solar system, that the processes which create them are the same throughout the universe, and that if life exists here, then it must also exist elsewhere.

The main problem, basically, is one of distance.

The nearest possible planetary system is located in Centaurus, the nearest star system to earth, some 4.3 light-years away, or approximately 26 million million miles. Radio signals would need over an eight-and-a-half-year time span between question and answer. For Tau Ceti, the question-and-answer time span is over twenty-two years. For stars hundreds or thousands of light-years distant, whole generations of men, and even civilizations, would pass between question and answer. Even a simple form of radio communication, not to mention physical travel, is at least for the present hopelessly impractical over great stellar distances.

Still, the search for the existence of intelligent life on these "other islands in space" is being undertaken. The first serious attempt at radio communication was Project Ozma, initiated in 1960 by Dr. Frank Drake of the U.S. National Radio Astronomy Observatory, West Virginia. An attempt was made to detect intelligently directed signals from two solar-type stars, Tau Ceti and Epsilon Eridani, in the 21-centimeter line of interstellar hydrogen. After 150 hours of observation, the project was terminated in May 1961, with no results. The main reason given was that the equipment was not at that stage sufficiently well developed, but financial considerations were probably also involved.

A further two-year project is to be undertaken in the near future, in a worldwide effort, and although success cannot be guaranteed, it is certain that if no experiments are made, results will be nonexistent.

Apart from normal radio communication in the 21-centimeter wavelength of interstellar hydrogen, optical methods are being investigated with great interest. The use of the emission line of interstellar hydrogen has been chosen because it has been thought that this is the most likely to be chosen by intelligent races attempting to contact equal or inferior technologies.

To judge from our own pattern of development, the next stage from optical astronomy is that of detecting optically unseeable objects in space, a task undertaken by radio telescopy, which utilizes the 21-centimeter line of interstellar hydrogen. Any race with a technology advanced sufficiently to undertake radio telescopy would be able to trace artificially directed signals among the background emissions of hydrogen.

The optical laser has provided sources of electromagnetic radiation more than 10,000 times as powerful as those obtainable with microwave generators. Pulsed powers of more than 1,000 megawatts have been attained, and much higher powers are not beyond our reach in the immediate future. (In 1962 a team of scientists bounced a laser beam off the surface of the moon.)

Another method of communication is a form of simplified television. The principle of a pattern of light and shade transformed into electrical impulses and resolved again into the pattern of light and shade (the telephoto image) is feasible for interstellar communication. A simplified transmission consisting of a series of ones and zeros, coded for translation to an image, could transmit understandable pictures of atomic structures, astronomical constellations, and a model of a planetary system—even a simplified image of a life form such as a human being—which would enable an intelligent receiver elsewhere in space to understand what we looked like and where we were located.

Even though we may spend years trying to communicate with other intelligences in space, we cannot *know* that such intelligences exist until we contact them. At the moment, it is virtually a matter of faith based on the assumption, from our present knowledge, that what has happened here has happened elsewhere.

On the other hand, the assumption that other intelligent races may exist within the universe is not entirely without foundation. Although the evidence is both slight and inconclusive, many curious incidents give at least some grounds for such a supposition.

Jupiter, for example, is a very powerful emitter of radio signals. Unlike general interstellar background radio noise, these "transmissions" are modulated in regular bursts at regular intervals of time. Since they occur when the satellite Io is in certain positions relative to Jupiter, it has been suggested that the satellite controls the radiation transmission of the planet. How a tiny satellite is able to do this, considering the vast bulk of Jupiter, is something of a mystery. The radiation bursts more closely resemble the sequence of narrow, frequency-modulated beams from a terrestrial radio beacon.

This is not to say that Jupiter is the home of an intelligent species that is transmitting signals for someone to pick up. But the transmissions may originate from a remotely controlled beacon, or an automated radio probe planted within the solar system from another solar system. Such an interstellar radio probe may have been placed in the Jupiter-Io system by means of a technology of communication devices beyond our present ability. Perhaps someday it may be possible to make sense of the pulse sequences (which may be a constantly repeated signal, beamed until someone cracks the code and replies).

Some forty-two years ago, Stormer and van der Pol reported unusually powerful long-delay echoes for which no source has been identified and no convincing explanation has been found. This onetime phenomenon may have originated in an interstellar probe located within the solar system.

The location of interstellar probes within the system would be difficult to detect since such mechanisms would be very small in relation to the size of the system. A likely place to search, however, would be the Lagrange nodes (named for Joseph Lagrange, 1736–1813)—five points in the vicinity of every pair of heavenly bodies which have zero gravity. Objects could rest in these areas without having to orbit a larger body. Such points would be ideal for interstellar probes, which could rest undisturbed for as long as necessary. Hungarian astronomer Zoltan Kopal has suggested these nodes as ideal places for space platforms (provided they are not already occupied!).

There may also exist channels of communication, or communications devices within our own system, which we are unable to detect.

We may already have been noticed by superior communities without our knowing it. UFOs and strange flying objects in earth's skies, observed even thousands of years ago, may be means of surveillance of our planet by other races. Such survey probes could be programed automata, electronic eyes and ears, with built-in safety devices to keep them out of our hands. This would explain their elusiveness from detection or capture by any of our devices—radar, aircraft, or missiles. They may have been examining us for a very long time, and, aware of our technical progress, would always be one jump ahead.

Most of these phenomena have been explained away, rather than explained. Weather balloons, unfamiliar aircraft, and the debris of artificial satellites did not exist a thousand years ago, yet the reports of such objects existed then, and are remarkably similar to modern accounts.

Why a probe lasting centuries should be undertaken without making contact with the inhabitants is a mystery. There may have been isolated contacts from time to time. The first chapter of Ezekiel has often been interpreted as a UFO contact, as has the Fatima miracle. More recent contacts always seem to happen in isolated places and with solitary individuals who carry no position of authority among our various governments.

Of course, we are looking at this from our own point of view. An extraterrestrial race, particularly if they are a long-lived species from a civilization with an extended time scale, may consider a survey lasting several thousand years to be a study of moderately short duration. If the survey probes are automated, they could be left to carry out their tasks indefinitely.

According to the ideas contained in this work, and drawing also upon certain aspects of mythology, *it seems likely that if the unexplained percentage of UFOs are real, and if they are vehicles of an extraterrestrial civilization, then that civilization they represent is a hominid one.* If the vehicles have live crews, it is more than likely that they are human, in general shape if not in detail.

If the human race of earth was of extraterrestrial origin, and arrived here some 40,000 years ago, *then the civilization from which they descended, the "mother culture," may still be in existence.* Although a span of 40,000 years is a great deal of time by earth standards—the average life of any given civilization has been about 3,000 years—we could expect much longer periods for the stable continuity of space-traveling communities, particularly interstellar communities. If the members of such communities were of long life span, such longevity could lead not only to an extension of the culture as a whole but also to greater stability over time.

Nor is it beyond the bounds of possibility that this world was surveyed by craft from other systems long before the human race took up occupancy. Alien ships may have drifted down through the skies of earth millions or tens of millions of years ago from civilizations or older stars, civilizations which have themselves been dust for thousands of centuries.

The foregoing is, of course, entirely hypothetical. The point is that the existence of other cultures in space is inferred from our knowledge of the physical processes of the universe, together with certain specific unexplained phenomena:

• *Dr. Gurlt's Cube:* This strange object was discovered in 1885 in a Silesian coal mine in a block of coal from the Tertiary epoch. Since it was inside coal from a seam being worked, it could not have been inserted into the coal, and could only have gotten there *before* the coal beds were laid down, *tens of millions of years ago.* It took the form of a 26-ounce cube 67 by 47 millimeters. Two opposite faces were rounded. A deep incision ran around the cube near its center.

Analysis showed its composition to resemble nickel-carbon steel. The sulphur content was far too low for it to have been a natural pyrite. Steel, we know, does not occur naturally, but is a manufactured substance. Most authorities who examined the object declare it must be of an artificial nature, but could not agree as to its origin. The object was in the Salzburg Museum until 1910, but it has not been seen since.

If this object had been buried for millions of years in coal, and was artificial, then it was not fashioned by human hands on earth. Was it junk, or a defective piece of equipment from a visiting spacecraft in the Tertiary epoch? Something thrown into a swamp, which in the course of time turned into a coal deposit?

• "*A bell-shaped vessel,* 4½ inches high, 6½ inches at the base, 2½ inches at the top, about 1/8 inch thick, resembling zinc in color, or a composition metal in which there is a considerable portion of silver. . . . This curious and unknown vessel was blown out of solid pudding stone, fifteen feet below the surface."

Was this vessel also artificial in nature, enclosed in the rock before its formation?

How many other objects may be discovered, preserved in rock or coal far below the earth's surface? Some may never be discovered, others may have disintegrated completely, and something may yet turn up, perhaps tomorrow, perhaps not for a hundred years, which we cannot deny is artificial.

As we have seen, one of the main problems of communication and contact between different cultures in the galaxy is distance. In the case of our sun, the stellar distances are of the order of one star per ten light-years; in the more densely populated regions of the galaxy, the ratio is roughly one to one. In the more densely populated regions of the galaxy, communication would be easier because of shorter distances between the stars, and consequently, their planetary systems.

Models of hypothetical galactic civilizations and their space-flight potential have been created by several authorities in physics, notably Dr. Lipp, von Hoerner, and Dr. S. S. Huang.

Von Hoerner says of galactic civilizations: "We expect to find a high activity in communication at shorter distances (200 to 300 parsecs) between civilizations of extremely long time scales, and very little if any activity at greater distances, from civilizations similar to our own."

A civilization of short time scales, or of one whose members are short-lived, would have extreme difficulty in communicating over even relatively shorter distances (up to 100 light-years). This is even more likely if we are bound to the speed of electromagnetic radiation for transmission purposes.

Carl Sagan of NASA has estimated that there is a good possibility that ten civilizations exist within the distance of 360 parsecs (1,000 light-years) and this would hold true for the galaxy as a whole; ten civilizations would exist over a diameter of any given 1,000 light-years at the same time.

Dr. S. S. Huang estimates that the number of habitable planetary systems is approximately 3 to 5 per cent, 5 to 8 billion habitable planets in the galaxy.

Based on these projections, the birth, progress, and eventual mystery of space technology by civilizations in a given area of space has been estimated at one every thousand years. Surveys of systems by intelligent races could be expected at this interval, although this does not necessarily mean that earth would be visited every thousand years. Oddly enough, however, we do seem to observe a thousand-year cycle of UFO phenomena, which may be related to this estimate, and certain religious writings often refer to the thousand-year visits of the "gods."

What could be UFO sightings were recorded in substantial numbers at the time of Ezekiel, roughly 1000 B.C.; another group at the time of the Roman Empire at the commencement of the Christian era; another in the period 1000 to 1200 A.D.; and yet another in the nineteenth and twentieth centuries—time intervals of roughly a thousand years each.

Models of intelligent life in the universe and percentages of space-traveling or space-communicating civilizations are based on mathematical *probabilities*, some optimistic, some pessimistic. These probabilities are also related to probability hypotheses for the development of life on earth and the evolution of the human species.

Is probability enough? It may seem that we are asking a great deal of the mere chance—even the fact of our own existence on earth. Chance mutations, working over tens of thousands of years, are said to have created intelligent human life from apelike ancestors, as well as thousands of other living species. It may seem to be stretching the laws of probability to the limits of credulity to expect the universe and its life forms to operate so uncertainly.

Perhaps there is more—much more—to all of this than mere probabilities. The rigid formula of the DNA blueprint, for example, or the regular mathematical precision of nuclear transformations, is hardly a matter of chance. Even Heisenberg's Uncertainty Principle in subatomic physics, which formulates a random factor in particle states, may be only a partial model of set patterns we have not yet discerned.

Perhaps, then, it is not all chance that civilizations exist in the galaxy, or that they arise in different parts of the galaxy. We have already remarked that it will eventually be necessary for the human race to leave earth if it wishes to survive. Scientists have recently suggested that the time is not far distant when at least a portion of the population will have to migrate to other planets. The situation which applies to earth may well have applied to other civilizations elsewhere in space, both in the distant past and continuing into the future.

Is it mere chance that such an "urge" may be built into the human organism as a survival factor? Or that this same factor applies to *all* intelligent forms in the universe?

The intelligence of an organism and its ability to use this intelligence to survive is in the last analysis a product of *mind.* And thought, or mind, may be a separate entity in its own right, fulfilling its purpose through the medium of the physical universe.

But what *is* the origin of intelligent physical organisms? Man could be the descendant of a space-traveling race. This solves the problem of man on earth without invoking either evolution or miraculous creation, but it does not solve the problem of *where* and *how* such an intelligence could have originated.

In *Life, Mind and Galaxies,* Dr. Axel Firsoff has said that mind itself may be a fundamental property of the universe, the universe itself may be alive, and the galaxy may be an intelligent entity of a high order, operating on nuclear rather than biological principles, on a level beyond our awareness and comprehension. Perhaps the reactions that created the galaxy commenced in the galactic nucleus, in the early stages of its formation, and the elementary particles of mental energy were diffused outward.

We could therefore postulate that an intelligent physical life form arose on a planet of a star near the galactic nucleus. How this form arose we are unable to say—perhaps there *was* some kind of "divine creation." Perhaps it arose only once, in human form (which seems remarkably well designed as a functional, machine-handling organism), or there may be other, nonhuman, intelligences.

Upon reaching a sufficient level of progress so that space travel could be commenced, this intelligent life form spread outward through the galaxy, traveling from suitable star to suitable star, and leaving some of its members on habitable planets or planets that could be made habitable, in the hope that they would take root and survive.

Possibly there is a never-ending shift and movement of populations throughout the galaxy, as races are compelled to move their homes when their stars become older and unable to support life within their own planetary systems. As we have noted, the sun will not always support human life. The time will come when no degree of adaptation by the human species will be sufficient as the sun, reaching critical temperatures, expands and the inner planets are engulfed.

Of course, life could always be transported to the outer planets as the sun becomes hotter. The atmospheres of planets like Jupiter and Saturn, which have a low density for their bulk, may eventually be vaporized away by the heat of the expanding sun. In the course of millions of years, they may become capable of supporting human life, either genetically adapted to the environment or protected in artificially created environments.

This would be only a stopgap, however. After the sun has reached maximum size, the cooling process will begin and the sun will shrink to become a red dwarf, once again freezing the outer planets. Before this stage is reached, those humans settled on such worlds will have to seek a new, younger sun, a G-type star on the main sequence, with a stable life of many millions of years.

On such a star they would have the time necessary to prepare themselves scientifically for a move to another planet when the necessity arose.

This concept of life following a pattern of migration through the galaxy is not particularly new. K. E. Tsiolkovsky said that *colonization, not evolution, may be the major factor in the spread of life in the universe.* If intelligent forms do migrate to new worlds, they may also take with them seeds, plants, and animals of various sorts to start basic agricultures. In the case of our planet this would explain why certain basic foodstuffs, grains and even flowers, and perhaps also some domestic cattle, have no traceable wild ancestry. Such ancestry may never have existed—*on earth.*

If we look at the universe and its life forms on the basis of this kind of long-term development, we may see that it is not so much chance that is operating, but the unfolding of a logical and ordered plan.

Perhaps this is the way it was, and is to be, and perhaps our original home is many times removed from this one.

5 WAS EARTH COLONIZED?

We are in the infancy of space flight, the "dugout canoe" stage. In the second half of the twentieth century, we have reached the moon in manned vehicles, and automatic probes have reached Venus and Mars. Clearly we have a long way to go even to explore the solar system.

Voyages to the moon can be accomplished in less than a week, but a voyage to Mars will take from two to three years with present propulsion systems, even though, astronomically speaking, Mars is almost on our doorstep, "only" 140 million miles away. With our present technology, by the time we are ready to launch probes to Saturn and Jupiter, these voyages will probably take about nine years. Plainly, present techniques are not going to allow us to explore even the solar system without enormous difficulties.

However, with the accelerating advance of technology, the fantastic of today is the commonplace of tomorrow. In the nineteen-eighties, when voyages by manned ships may be undertaken to Mars, yet-to-be-developed propulsion systems may have reduced the time factor to a few months. Tests have been conducted recently that indicate that nuclear-propulsion systems are feasible, and that sustained high velocities can be achieved within a few years, possibly without the need for the extravagant bulk of fuel which makes rockets so enormous and unwieldly at the liftoff.

One problem of interplanetary flight—escaping earth's gravitational field—involves almost all of a space vehicle's fuel capacity. Building the ship in space in earth orbit, at a distance of several thousand miles, would eliminate or greatly minimize this problem.

Newly developed miniature atomic power units will enable robot spacecraft to travel to Jupiter, Saturn, Uranus, Neptune, and Pluto and transmit information back to earth. NASA announced in September 1970 that it hopes to launch the first of such probes toward the end of the nineteen-seventies.

Voyages to the more distant planets of this system, distances of up to 2.7 billion miles, are expected to last nine years. The nearest star system, Centaurus, is some 26 trillion miles away. With the anticipated velocities for interplanetary craft, when used for interstellar travel, a voyage to Centaurus would take around 80,000 years. Even if the velocities were doubled or trebled, such voyages would last thousands of years.

Unmanned probes obviously present fewer problems than manned ships. Nuclear power units could enable the craft to run for thousands of years, if necessary. Automated probes could be programed for many complicated tasks, including the search for life-bearing planets of other stars. Once again, however, time may render such probes—at least, to very distant destinations—virtually useless. Even though the information they transmitted would take only four years or so to reach earth from the vicinity of Centaurus, the probes themselves might take thousands of years to reach that system. The civilization that launched them may well have passed away, or earth reverted to barbarism in the intervening period. On the other hand, technological advances may make it possible to overtake such probes with manned ships before they have completed a quarter of their journey.

The minimum requirement for interstellar flight is a propulsion system which provides speeds near the speed of light. Theoretically, such systems exist; they must be tested to see if they are actually workable. *Nuclear engines,* which would develop continual thrust to build up velocities approaching the speed of light during the period of the voyage, are one possibility.

Another is the *ionic drive,* which operates on the principle of discharging a stream of electrons in much the same way that an electron stream is emitted in a cathode-ray tube. Velocities up to 90 per cent the speed of light are anticipated by this method.

A similar propulsion system using light particles (the *photon drive)* is also theoretically possible and would achieve velocities comparable to those of the ionic drive. It is thought that these two systems will operate most effectively where gravitic influence is negligible, in interstellar space. Within the solar system, they may not be particularly effective. To surmount this obstacle, several systems could be used: a nuclear-powered propulsion system within areas of powerful magnetic influence, and the ionic/photon drive in interstellar space.

The outlook may seem pessimistic at the present time, but we must remember that in less than a hundred years we have progressed from almost a walking pace to considering building engines which approach light speeds. As technological progress tends to snowball, we can reasonably expect a hundredfold progress in the hundred years to come.

According to our present knowledge of physics and Einstein's theory of relativity, speeds exceeding the velocity of light are not possible. Mass appears to increase with velocity and, according to Einstein, at the speed of light mass becomes infinite and therefore ceases to exist.

At velocities near the speed of light, there is, according to our understanding of Einstein's general theory of relativity, a very interesting effect, which would be both a help and a hindrance in interstellar flight. This effect, known as "time dilation," has been provisionally demonstrated in laboratory experiments.

Two batches of radioactive substance were spun at many thousands of revolutions per minute, one faster than the other, simulating two different velocity rates. A radium clock counted the emission of particles, tracking the decay rate, and it was found that the material spun at the faster rate (the one that was traveling faster), had a slower decay rate than the material rotated more slowly. Apparently the faster material was "dying" slower; time, to it, was relatively slower.

Simply stated, the time-dilation factor exerts the following effect: as an object approaches the speed of light, time becomes slower (for the object). For a spaceship and its crew, traveling to Tau Ceti at 97 per cent of the velocity of light—a voyage of some 22 to 23 years in relation to earth—the elapsed time would be a matter of *weeks*. From the point of view of the crew, where elapsed time is concerned, stellar voyages at velocities near the speed of light compare favorably to interplanetary flights at much lower speeds. A flight to Mars lasting six months, for example, would be much more of a strain on the crew than a flight at velocities near the speed of light to Tau Ceti.

Admittedly, the psychological disadvantages of such voyages would be great. While the crew have aged a matter of weeks, all their relatives and friends on earth will have aged over twenty years. It would probably be more desirable for crew members on interstellar flights to have no close relationships on earth.

For long-distance flights lasting a hundred or more light-years, the problems are even greater. Several centuries will have passed on earth, and great changes will have taken place during the absence of the crew. It would be as if a man left on a voyage during the eighteenth century and returned to the totally unfamiliar world of the twentieth century.

On the other hand, time dilation ensures that personnel of space vehicles traveling at such velocities will reach distant destinations without suffering the effects of extreme old age. In all probability, however, *there would be no coming back,* particularly for voyages of great distances. For this reason such voyages would be undertaken by a group sent to settle a newly discovered planet. Interstellar business, trade, and cultural travel would be hopelessly impractical.

Vehicles traveling on interstellar flights at much lower speeds (several thousand miles per second) would make voyages lasting for centuries. What was once science fiction has become a matter of serious practical consideration.

There are two popular concepts regarding such long-duration flight "arks":

• The *self-contained artificial world* of great size in which generations of men and women would travel until their descendants reached their destination. The success of such a model would depend less on the practicability of the technology than on the ability of the human organism to cope with a closed environment and a limited population over a long period of time.

The purely mechanical problems are not insoluble. Waste products and water could be recycled and would probably be adequate for long voyages (the moon flights would seem to point to the fact that the problem may not be a shortage of air and water, but an excess). Foods could be grown in hydroponic gardens, for it would be impossible to carry sufficient stocks of prepacked foods, even in concentrated form, for a voyage lasting centuries.

In a sense, earth is a spaceship as it travels through the cosmos, and is a closed environment, relying on recycling techniques on a large scale. Our interstellar ark would have to be fashioned on similar principles.

Consider, for example, the problem of meat, milk and milk products, clothing, and various other by-products supplied by domesticated animals. A number of cattle and sheep would provide not only these products but also a nucleus of future herds and flocks upon arrival on the new planet. Vegetation grown in hydroponic gardens would not only provide essential foods, but also assist in regenerating the atmosphere of the ship by oxidizing it.

The birth rates of both the human and the animal populations would have to be rigidly controlled to prevent overcrowding. Those who died on the voyage would be neither buried nor cremated; their tissues would be converted to usable materials. Birth would be geared to death; babies would be permitted only to replace someone who died, so that the numbers could be kept in balance.

It is possible that after many generations the people on the ship might forget the purpose of the voyage, might not even remember there ever was a purpose, either to the voyage or to the ship, and in the end the ship might become the world, the be-all and end-all of existence. (See Clifford Simak's excellent treatment of this theme in his story "Target Generation.")

This may not matter to a great extent. The functioning of the ship—flight path, velocity, monitoring systems—could be controlled from a computerized control room kept securely locked for the duration of the voyage to prevent interference with its function. Even monitoring and landing on suitable worlds when found could be done automatically from programed information fed into computer banks.

Upon landing, the ship could automatically be opened, and devices put into operation to ensure that the people left the ship. After generations in space, with no concept of a natural and frightening world, they might be reluctant to leave their artificial womb, and might have to be forced to leave it.

Problems would arise from two sources: genetics and human

nature. The tendency of a small confined population to inbreed would be minimized by careful selection of personnel, from the standpoint of both personal health and health of family ancestry.

Human nature poses another serious problem, particularly if the personnel develop a type of class-conscious society that would lead to eventual friction, and perhaps conflict, where all or most might be killed. The solutions to both of these problems are probably to be found by psychologists rather than by technicians.

• The second approach to the problem of long-duration space flight involves *deep-freezing techniques,* presently in their infancy. Lowering of body temperature for surgery has already been carried out with some success on patients who are, because of respiratory difficulties, unable to take standard anesthesia. Deep-freezing could produce a state of suspended animation, suspending all bodily function to the point of apparent death, with the patient being restored to active life by being thawed out.

By the use of this technique, a fully automated vehicle could be launched toward a distant destination. Its occupants, frozen in suspended animation, could be automatically revived at the end of the voyage. The ship's flight and functions could be automatically controlled for the entire voyage, or selected members of the crew could be awakened at intervals of time to check the condition of the ship, its mechanisms, etc.

This method obviates the necessity to carry major life-support systems, and also ensures that the original members of the crew arrive at their destination with virtually no effects of aging. There would also be none of the psychological problems experienced by ship-bound descendants of an original crew.

Whether there can ever be such things as FTL (faster-than-light) drives, or space warps (which cut across the space/time continuum, annihilating distance), is a problem for the future. Faster-than-light drives may be feasible, provided that the Einsteinian concept of the limitations of electromagnetic variation for velocity is wrong. The temporal effects of such a drive cannot be guessed. It may mean virtually instantaneous travel, or even temporal reversal, a sort of traveling backward in time.

Space warps involve using extradimensional states, quite beyond the bounds of any science we can so far envisage. However, it is not wise to say such and such *cannot* be done, for we are doing things today that to our ancestors of a thousand years ago seemed absolutely impossible.

Other solutions to the problems of interstellar travel may take us into the realms of psionic (ESP) phenomena and metaphysics:

• Could communication be effected across interstellar gulfs by *telepathy,* if it could be proved that such a communication form exists? Would telepathic communication suffer from the limitations of electromagnetic radiation, or does it operate independently of the limitations of the physical universe?

• Would it be possible to effect physical transfer by *teleportation* (transmitting physical objects from place to place by the power of the mind)? Would the same limitations apply here, or would transfer be instantaneous?

We have suggested that humanity could have been of extraterrestrial origin. If we allow such an idea, can we now show any evidence for such an event? Could there perhaps be physical remains somewhere, or a trace of the technology involved, either in physical form, or in the memory of the race?

It may be easier to look at this problem in reverse, by postulating how we may deal with such an interstellar colonization project in the future. As we have said, the vehicle, or "ark," could be controlled automatically. Upon landing, one of two things would happen to it. Since such vehicles are extremely expensive and useful items of equipment, they could be programed to return to the solar system for further groups of colonists. Or, if the distances are too great and the time lag too excessive, the vehicle could be used first as a temporary shelter upon the new world, and then broken up and its parts used for other purposes. Its power plant could be used to power the nucleus of the first settlement, its structure for the construction of new buildings.

In the course of time, the ship would cease to exist and even its parts become useless, worn out, discarded to rot away. Within several hundred years, no trace would remain. In the course of time, the descendants of the original colonists, having no physical proof of how they landed, might deny that this had ever happened.

Such colonists might even have forgotten their point of origin by the time they landed, or carry only myths of a long-vanished paradise world in the sky. Extant mythologies from many parts of the world contain such allusions to a paradise beyond earth—"heaven"—which could well have basis in reality.

If earth *was* colonized by an advanced culture from elsewhere in the universe, why did not their gadgetry and their knowledge survive even to the present? There are two answers to this, which will be dealt with at greater length later.

1. The devices could wear out and not be replaced, and the knowledge could be lost in the course of time.

2. The colonists, as their numbers grew, might suffer wars and internal strife, losing their knowledge and reverting to barbarism. Valued members of the colony might also be killed by wild animals or in accidents. The possibility that such a regression may have occurred on earth in the past will be one of our major concerns.

Regarding technology, again, it is of interest to look at the problem in reverse. The colonists of our future voyages are on a one-way trip: there is no return and no contact with the home culture, because of the distances involved and the time factor. They are thrust into a new, raw, possibly harsh and unfriendly environment, to begin the struggle to tame a wilderness. No explorer in Africa in the past was so isolated from civilization. Many hazards will face our future colonists: climatic extremes, poisonous plants, earthquakes, floods, dangerous animals.

What kinds of weapons would be most useful to a small group of humans on a wild, uninhabited planet? They would probably need projectile weapons. Perhaps our colonists should include rifles and a supply of ammunition among their supplies on the ship. But no matter how much ammunition they brought with them, eventually the supply would be exhausted. A rifle and its shell are products of a highly sophisticated technology. Even a hundred years ago, firearms were not particularly efficient or accurate. The famous Colt .45, for example, was a hopelessly inaccurate weapon beyond a twenty-yard range, and was as likely to blow up in the user's hand as it was to fire.

The point is, the supply of ammunition would probably be exhausted along before the colonists had been able to develop their technology to the point of being able to produce ammunition for their rifles. In any case, in their development of the new world, rifle bullets would come pretty low on the list of priorities. And a rifle minus ammunition is only a moderately useful club, no more so than a hefty chunk of a good hardwood.

The bow and arrow, on the other hand, is an extremely effective weapon and for hunting game it is more useful than a gun. It is silent, and so does not frighten the rest of a herd or flock away. Once used, whether it has found a target or not, the arrow can be reused, an advantage rifle ammunition does not have. Most important, the bow and arrow is simple. It can be made from a variety of materials found in nature, and a fire-hardened point, or a sharp stone arrowhead, will kill just as effectively as a finely machined steel one.

What our colonists on a new, uninhabited world need, then, are simple things and mastery of simple processes—the ability to make and use the backstrap loom for the manufacture of cloth, pottery-making, candle-making. This is the stuff of which survival and civilization are built, rather than a host of technical gadgets, which rapidly become worthless. The technology can come later, when the world is made more habitable, with the knowledge they have brought with them—provided they have not forgotten it.

If we look at things in this light, the refinements of a space-traveling technology as the earliest records of our ancestors would not necessarily exist. Furthermore, we do not know why this planet may have been colonized. It may have been planned or it may have been by accident, when a ship or ships landed here in an emergency and were unable to leave again. Perhaps earth was originally a military base or outpost, or a communications station. It could be that those who came here were fleeing from some catastrophe, or that they were criminals or undesirables, expelled from more civilized societies.

They may have arrived here with simple tools, or with the products of technology advanced so far beyond our present technology that we would not recognize any remaining artifacts if we found them. Some of our most sophisticated gadgetry would make no sense to a primitive. What, for example, could they make of a radio transistor, or a cathode-ray tube, or a printed circuit? Not only would they lack meaning, they would be totally useless. A laser gun, or the remains of one, would be equally puzzling to a Roman legionary.

At this point we can only infer that man has an extraterrestrial origin. *If man has suddenly appeared on earth, complete with his intelligence and certain basic knowledge, and if his origin cannot be traced from any other terrestrial form, then he must have come as he was, from somewhere. If that somewhere cannot be traced to a point on the earth, then it must be a point away from the earth.*

We have found evidence that certain Pacific islands were uninhabited until the eleventh century A.D. Evidence—buried weapons, cooking utensils—shows habitation from the present time back to approximately this period, and then nothing. Some say that these

Polynesians migrated from the Asian mainland, or Malaysia or Burma; others, that they hailed from South America. No one suggests that they evolved or were miraculously created in isolation on the island.

Perhaps the circumstances which apply to the Polynesian islanders apply to the human race as a whole; that is, *true humans appeared on the earth at a given point in time, with clothes, fire, weapons, shelter, and with a native ability and intelligence that is certainly no less than that possessed by peoples of the present time.*

The concept of the evolution of man could be utterly false.

We hope to have shown, thus far, several things:

• That the universe comprises star and galactic systems composed of the same materials and developed by the same processes in all its parts, and that the same principles of thermonuclear reactions apply, whether to our sun or to an unnamed star in the farthest galaxy we can see through our telescopes.

• That life, being composed of elements common throughout interstellar space, will follow the same laws whether it be on earth or on some distant planet, and that "will," "thought," "mind," and "intelligence" may be universal properties operating within energy fields through as yet unknown laws.

• That the behavior patterns and urges built into the human organism by the mysterious agency of mind includes the urge to explore the universe, and that this "urge" exists wherever intelligent beings exist.

• That space travel is not merely a possibility, but has already begun.

• That extraterrestrial visitation—and hence, extraterrestrial colonization—of the earth is a scientifically recognized possibility.

Let us now turn our attention to man on earth.

The evolutionary sequence for the development of life on earth has been hypothesized by scientists, starting with the probiotic soup (early warm seas), to plants and fishes, amphibians, giant reptiles, early mammals, and the appearance of man. Neat geological epochs have been theorized—with a few admitted holes here and there, but on the whole a reasonable picture.

The three main epochs are Paleozoic (early life), Mesozoic (middle life), and Cenozoic (recent life). Somewhat surprisingly, the ancient periods are dated a great deal more confidently than are more recent times. With the onset of the Cenozoic period, the sequence of things begins to get confused and by the period known as the Pleistocene, dating becomes thoroughly unreliable.

The Cenozoic, estimated to have commenced some 70 million years ago, is subdivided as follows:

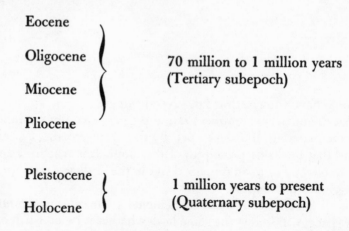

Eocene

Oligocene

Miocene

Pliocene

70 million to 1 million years
(Tertiary subepoch)

Pleistocene

Holocene

1 million years to present
(Quaternary subepoch)

During the Cenozoic period, the great reptiles were replaced by mammalian forms. Why the reptiles, masters of the world for hundreds of millions of years, should have disappeared, replaced by the mammals, is an unsolved mystery. Mammals' ability to survive greater variations of external temperature because they can maintain a constant internal temperature may have a bearing on the problem.

This is, however, only a partial, unsatisfactory answer, because a great number of species of cold-blooded forms (reptiles) survive to this day, many of them in Europe, which has extremely cold winters.

Some scientists suggest that radiation from a near supernova may have killed the reptiles. Why did it not also kill the mammals and insects, which latter seemed to have survived, some almost unchanged, since the Paleozoic?

The dating periods of the Cenozoic are only approximations, and are open to change. Such changes in dating have been frequent and often drastic, especially within the last million years, the period of the Pleistocene. The dates are so vague and contradictory that almost any guess is as good as any other.

The Ice Age, it is thought, started roughly 1 million years ago, at the beginning of the Pleistocene (most recent) period, and ended some 10,000 years ago, when the last of the ice melted in temperate latitudes. The Ice Age was not a continuous age, but four periods of glaciation separated by three long interglacial periods.

The date assigned as the ending of the last glaciation, or interglacial period (known in Europe as the Wurm), has been subject to great fluctuations. At first it was thought to have ended some 30,000 years ago, then the figure was reduced to 20,000 years. Now the figure has been brought down to 10,000 years ago, although some scientists place it as recently as 8,000 to 6,000 years ago. These, it must be admitted, are great variations over a comparatively short period of time.

The periods of the Pleistocene Ice Ages, their causes, when they happened, or whether they happened, will be discussed later in this book. What concerns us here is the evolutionary theory according to which the precursors of humanity—the hominid types of anthropoids—arose during the past million years.

Estimates of the original age of the first hominids vary from 1 million years to many millions of years for the ancestral types from which the hominids were descended. There also has been some question as to whether some of these ape-men were actually ancestors of man, or whether they were anthropoid apes on the road to humanity that became extinct.

During the long warm periods (which go back some 70 million years) before the Pleistocene glaciations, many mammalian forms flourished, including many kinds of apes and monkeys. Lemurs, it is supposed, are among the most ancient of all apelike forms, with a history of some 100 million years or more.

This dating would place certain primitive forms—lemurs and also tarsiers—just before the end of the Age of Reptiles. There was a great expansion of both pongoids (monkeys) and primates (apes) during the Miocene and Pliocene. The chimpanzee, which appeared during the latter part of the Pliocene, although it resembles man in appearance and intelligence more closely than any present-day anthropoid, is apparently not directly related to man.

What are regarded as the likely ancestors of *Homo sapiens* arose during the last million years—a period during which large areas of the earth are reckoned to have been frozen by the glaciations. Among the names most frequently mentioned are the Australopithecus, roughly a million years ago, Java (Trinil) man 500,000 years ago, and *Homo pekinensis*, some 300,000 years ago. The varieties of primitive man and near-man, among which are Neanderthal man, Heidelberg man, Grimaldi, and Swanscombe, are dated some 35,000 to 75,000 years ago.

Why did the hominid primates appear only during the last million years and not during the long warm periods of the Miocene and Pliocene? A great many apelike forms flourished during this time; according to evolutionary theory, it would have been an ideal time for the precursors of humanity to have appeared. The rapid evolutionary development that led to man should therefore have taken place many millions of years ago. This did not happen. *True man,* Homo sapiens, *cannot be traced back further than 40,000 years ago,* and much of the evidence in the form of skeletal remains may have to be placed at a period much closer to the present.

Even among anthropologists there is disagreement as to whether Dryopithecus, an ape that lived in the Miocene (25 million years ago), was an ancestor of man. This species disappeared some 9 million years ago, leaving a gap of 7 to 8 million years between its disappearance and the appearance of the Australopithecines just over a million years ago. The remains of Ramapithecus, a small apelike creature with an almost human face, were discovered in the Siwalik Hills in northwest India. There is, however, an even greater time span between this creature, whose appearance is dated at 12 million years ago, and the Australopithecines.

What is actually known of these gibbonlike apes of millions of years ago? Very little. From the discovery of part of a skull, or a few finger bones, or part of a lower jaw, the size and shape of the creature is deduced. If these creatures died out completely, where is the connection between them and the next apparent forerunners of man, the Australopithecines, or Java man? There is none. Were the Australopithecines as manlike as has been supposed? Measurements of the remains—various jaw and skull fragments—indicate that the brain was only a third as large as man's. The structure of this brain may have been no larger or more complex than that of a present-day dog.

It is claimed that the Australopithecines were toolmakers, that they used sticks and stones to hunt game, but is there any proof that they ever hunted game? All modern species of apes and monkeys are vegetarians. There is no reason to suppose that these early apes of the Miocene, Pliocene, and even the Pleistocene were not the same in their eating habits, just as their contemporaries, the ancestors of the horse, deer, and elephant, were vegetarians. There is no evidence that these early apes were meat-eaters.

Remains of other animals found near the remains of Australopithecus suggest not that Australopithecus was carnivorous but that it may have been the victim of carnivores. According to anthropologist Dr. Oakley, "The accumulations of nonaustralopithecine bones found in the australopithecine deposits can be traced to the activities of carnivores, including hyenas." It is more likely that the Australopithecines, like modern apes, were vegetarians and were hunted as food by larger carnivores.

Anthropologist Washburn concurs: "The australopithecines were themselves the game, rather than the hunters." And R. L. Lehrman, in his book *The Long Road to Man*, states: "Australopithecus was merely an upright, intelligent ape, not a man. The small braincase, bearing heavy ridges over the eyes, down the back, and down the centre, was more like that of an ape."

The next supposed human ancestor, *Homo erectus*, includes Java man, 500,000 years ago, and Peking man, 300,000 years ago. There is a great deal of uncertainty regarding even these types. Doubt has been cast on the theory that these creatures made fire; anthropologists now say their association with fire may have been accidental.

There are further anomalies with regard to accepted theories on these extremely ancient apelike forms. All have been found in the Old World. Nowhere in the Americas is there any trace of archaic hominid species. This is true in spite of the fact that the antiquity of true man in the Americas is probably equal to that of the Old World.

Migration theories, according to which true man migrated to the Americas from the Old World, do not account for the equally reasonable possibility that *Homo sapiens* could have appeared first in the Americas, and then migrated to the Old World. Nor do they account for the fact that Africa, which claims the oldest forms of "man-apes," is devoid of *Homo sapiens* remains until comparatively recent times, at a period after the termination of the Wurm glaciation.

If our dating techniques have any validity, Europe and the Middle East show evidence of the oldest occupation of true man, and no traces of pre-*sapiens* hominids. According to evolutionary theory, surely Africa and China (home of Peking man and also scarce in antique *sapiens* remains) should show the most ancient traces of *Homo sapiens*. The only explanation offered for this odd state of affairs is that true man migrated.

Why, however, should he migrate from warmer to colder climatic zones, where the problems of survival are so much greater? During the primary phase of *Homo sapiens'* development, he would have lacked the knowledge to clothe himself adequately and build adequate shelters. How would he survive the arctic rigors of Ice Age conditions? By the time he had learned these lessons, he would have died from exposure. According to the theory, however, this did not happen. There is no evidence that the first *Homo sapiens* was covered with fur.

Virtually all mammals, whether they live in tropical or arctic climates, are fur-covered; apart from a layer of fat beneath the skin in arctic-dwelling animals, the main difference in their coats is color. The polar bear is white; we know that the fox and the snow leopard change to a white fur in winter or in the arctic regions. This is not so much climatic protection as *protective coloring*.

Homo sapiens is virtually hairless, especially as far as body hair is concerned. Of the human races which inhabit the colder zones, one, the Caucasoid, is relatively more hairy than the Mongoloid, which frequently inhabits colder regions than the Caucasoid. This does not make sense in terms of either evolution or adaptation to environment. Man could not rely on the natural protection of a fur coat he never possessed. He must have protected himself artificially from the elements.

The discovery of Java man, a pre-*sapiens* hominid, included a skullcap, a left thigh bone, and three teeth. A fragment of a lower jaw was found in another part of the island in the same geological layer. Even if all these parts belonged to the same species, there is no guarantee that once assembled they would present a true picture of the creature. We have no way of knowing whether one of the parts belonged to a smaller, perhaps female, member of the species, and the others to a large male. Furthermore, remains of a true *Homo sapiens* of modern type were found in the same geological layer. Either true humans are as ancient as the ape-men, which does not support the evolutionary theory, or the Java man remains are merely those of a species of recent ape which has become extinct.

A fossil tooth found in Nebraska in 1922, it was claimed, belonged to a species of extinct anthropoid ape (the first found in the Americas). It was later found to have belonged to a fossil peccary.

When the famous (now notorious) Piltdown skull was "discovered," it was claimed that this was the famous "missing link" between ape and man, so desperately sought by anthropologists. Further investigation showed it to be a complete fraud, partly modern human, with cleverly doctored chimpanzee additions.

There is only a superficial physical resemblance between the anthropoid apes and man. In point of fact, the domesticated pig shows a greater physiological resemblance to man in the structure of its internal organs, and the bear not only walks very much more uprightly than any ape, but it has the wide-ranging food habits of man. From a purely logical point of view, both of these animals would seem to have as much claim to be human ancestors as apes have had.

The brain represents the greatest difference between man and all other animals. According to one explanation for the increase in size of the human brain, which apparently occurred as man evolved, man assumed an upright stance, and his hands were free to experiment with the handling of objects and later the fashioning of tools and weapons. This learning experience, the theory holds, assisted in the development of the brain. The brain's gradual development increased the usefulness of the hands, which led to a further development of the brain, and so on.

This theory, simple and straightforward as it appears, is riddled with difficulties. Recent experimental studies of chimpanzees in the wild reveal that they will use a stick to get at inaccessible supplies of food (for example, grubs in a hole in a tree trunk). If the stick is too long, they will break it and snap off any superfluous twigs, so that it is more fitted for the task at hand. By modifying a natural object, they have created a tool. Any of the intelligent apes in antiquity probably did a similar thing (one of them must have, according to evolution, as the first step on the road to humanity).

Why, then, has the chimpanzee progressed no further? The chimpanzee has been in existence for many millions of years more than the supposed ancestors of man, so it should have progressed further up the evolutionary ladder. As Erich von Däniken put it in his book *Return to the Stars,* according to natural evolutionary theory, why, by now, do chimpanzees not wear trousers? For the simple reason that all the anthropoids, living and extinct, *never possessed the capability to evolve into a human being*.

If we regard Peking man as one of the earliest ancestors of man, we have a time span of some 300,000 years until we reach our present state of development. According to the time scales of evolutionary development, this is a minute period of time. The development of the human brain in evolutionary terms has been described, at the same time, as literally an explosion of size and capacity. Such a vast increase in brain size, complexity, and ability is in fact totally at variance with the general evolutionary picture.

The concept of the evolution of *Homo sapiens* from an apelike ancestor is highly dubious. The entire essence and basis of the evolutionary theory is contained in one word: *mutation*. Evolution is survival through natural selection or favorable mutations.

A mutation is a change or alteration of a characteristic, or group of characteristics, in a particular creature. It is, generally speaking, an accident. As far as we can judge from present living forms, mutations are generally far from beneficial. Even if the mutated specimen survives to reproduce, the new generation will probably revert to the existing type.

Mutations can be caused by drugs (e.g., thalidomide), or by hard radiation. Deformed children born to women who have taken drugs or been X-rayed during pregnancy, however, can, when adult, produce normal children. (In the case of thalidomide children, unless there is any residual effect of the drug, we could expect them to produce normal offspring.)

Fruit flies subjected to radiation exhibit many different degrees of mutation, all of which so far would make the insect in a wild state unfit for survival. Even those which do not inhibit the survival of the creatures are certainly not an improvement on the original. With subsequent generations, the mutation, favorable or not, is bred out and the creature reverts to type.

Even if a favorable mutation—an increase in intellectual ability, for instance—had occurred to, say, an individual Peking man, it would appear from our present experience that it is extremely unlikely that such a mutation would be passed on to future generations. What happens to present-day species will have applied to species in the past.

One further point is little stressed by evolutionary anthropologists in connection with mutation. During the period we are concerned with, we have to assume that the hominid in question is virtually on a level with the animal world; the "mutation" which is to make him more human has not yet occurred. A mutant, whether favorable or unfavorable, stands a slim chance of survival. It is well known that in the wild, deformed young are usually either killed by the parents or left to die, as aberrants are not recognized as the young of the same species. This instinctive reaction is nature's rather harsh way of ensuring that only that which breeds true and therefore stands a chance in the struggle for existence shall survive. This is why in a wild state one almost never comes across an imperfect specimen, unless it is due to an injury. Imperfections are not permitted to survive.

If ancient hominid anthropoids lived in groups or colonies with a close-knit social order, like many species of present-day apes and monkeys, it is likely that any hominid which showed traces of aberrant behavior, even if it did not display any special or unusual physical characteristics, would no doubt get short shrift from the leader of the group. If not killed, it would be driven away from the group, where its chance of survival as a solitary animal would probably be slight. Even if it managed to become the pack leader by reason of greater strength, agility, or intelligence, it is extremely unlikely that its ability would be passed onto another generation.

Whether the mutation is favorable or unfavorable, species will revert to the norm structure laid down in the genetic pattern of the family, or phylus. Furthermore, the odds that two favorable mutations would occur at the same time, in the same quantity, one in a male and the other in a female, in one particular group of what are reckoned to have been relatively small groups widely scattered across whole continents, are so small as to be virtually impossible. Yet we are asked to believe that this is the way *Homo sapiens* developed from apelike ancestors.

But the situation is yet more complicated. Can we account for the development of *three* races of man (remembering that the odds against any "surviving" mutant species are astronomical, and that the odds against a "favorable" mutation are even greater)? According to evolutionary theory, for example, Peking man may have produced the mutation that led to the Mongoloid, Java man the Negroid, and either Heidelberg or Steinheim man the Caucasoid (white) race. These mutations would all have to have been favorable, and all of exactly the same kind.

In fact, these species are separated by great gulfs of time and distance—yet all these "mutations" were supposed to lead to three human species, all showing the same advanced characteristics. The odds are hopelessly against such unlikely combinations over periods of literally *billions* of years—yet we are expected to believe that all this could have happened *in less than half a million years*.

Furthermore, we have no evidence that widely diverse pre-*sapiens* hominids could interbreed with each other. All true humans, whether they be pygmies from the Ituri Forest or the most highly educated men, *can* interbreed, for they have a common genetic heritage, and at some remote period, a common ancestor. On the other hand, even fairly close families of modern anthropoids *cannot* interbreed.

The same rules in all probability applied to the "ape-men" of the past, even if we assume that they lived at the same period in the same localities. Unfortunately they did not, and it would seem extremely unlikely that these different types of hominid anthropoids ever met, let alone mated.

The process of adaptation has often been hailed as proof of the validity of the evolutionary concept. It is nothing of the kind. For example, the peppered moth of Britain has, within the last hundred years, darkened its coloring considerably. This adaptation has made the moth better able to blend with the darker color of walls and buildings brought about by industrial pollution, which in turn assures it better protection against its predators.

Very few creatures within the same phylus are able to crossbreed success-fully. The lion and the tiger can, and on occasion, do. This proves only that they may be variants of a common related type in the past, since dif-ferentiated because of environmental adaptation.

Is this an example of the evolution of the moth? It is not. The moth is still a moth, it is not evolving into anything else. There is no sign that it is becoming a higher form of life. What has happened is that the moth has adopted a protective color scheme in keeping with the changed conditions of its environment, in the same way that the fox changes to a white coat in winter.

Many anthropologists, who realize that traditional views of a direct transition from ape to man are largely untenable, are now suggesting that the various apelike forms are "experimental types," and from these earlier models, *Homo sapiens* developed as a separate species in some as yet not identified area of the world.

But no matter how many books are consulted on the problem of human evolution, we come ultimately to a simple statement which may as well have been said at the beginning of the book: *And then true man,* Homo sapiens, *appeared.* Where and how and when, we have no way of knowing.

Homo sapiens *appears, suddenly, out of the blue, complete, intelli-gent, with clothing, weapons, fire. With a curious mythology that owes nothing to an animal ancestor, and with an ability that, at the begin-ning, is at least equal to that of today.*

7

THE MYTH OF THE
"APE-MAN"

In the preceding chapter we hope to have shown that the evolutionary concept is scarcely tenable on either logical or scientific grounds. It is as much a matter of faith as the concept of divine creation and, from a strictly logical point of view, just as improbable.

The only thing that can be said with any certainty is that man appeared at a time which cannot be determined with any certainty at all, although most scientists are agreed that it was about 35,000 to 40,000 years ago.

We have some evidence of man during the past approximate 8,000 to 10,000 years, the period known as the postglacial, and increasingly less evidence the further back we travel in time. There is also considerable confusion and controversy as to whether certain remains stem from glacial or postglacial times.

Science has divided *Homo sapiens* history into archeological divisions, as follows:

Classification	Date
Historic	Present—A.D. 1
Iron Age	A.D. 1–1000 B.C.
Bronze Age	1000–2000 B.C.
Neolithic	2000–4000 B.C.
Mesolithic	4000–8000 B.C.
Late Upper Paleolithic	8000–12,000 B.C.

These classifications are generalizations, and the dates are tentative; in fact, the value of such artificial divisions is diminishing the more we learn of early man. Prior to the Bronze Age, for example, such a table implies that all previous proofs are variants of the period generally known as the Stone Age—a very misleading picture.

At present, civilization exists on many levels in different parts of the world, ranging from the highly sophisticated technical societies of the Western world, the Soviet Union, Japan, and elsewhere to the "Stone Age" cultures of the Australian aborigines, the native inhabitants of New Guinea, the Amazon interior, and the Bushmen of the Kalahari desert of Africa. In the twentieth century, in other words, we are still in the Stone Age in some parts of the world.

Five or ten thousand years from now—by which time our civilization may have long been destroyed by, say, nuclear war—who is to say what archeologists may discover? Perhaps all that survives such a holocaust will be the Stone Age tools of the aborigines and Bushmen, with scarcely a trace of our technical accomplishments. A few twisted artifacts or building foundations may remain, but in general they may conclude that the twentieth century was a Stone Age before civilization developed.

Characterization of the Bronze Age as the "first" age of metals is also misleading. Recent discoveries in various places indicate that copper and bronze were used long before this period—in certain cases, thousands of years before, which places them well within the Neolithic or Mesolithic periods. While these periods are useful pegs to hang things on, they cannot be taken too literally.

Homo sapiens' appearance has been dated to the middle of the last great glaciation, the Würm, which ended 10,000 years ago or less.

The earliest traces of true man occur in Europe, particularly Britain, Scandinavia, France, and Germany. *This, according to present theories, places their arrival in the intense cold of an arctic environment in the middle of the glaciation:* a climate of long bitter winters with a great deal of ice and snow, and wet, cool summers with temperatures barely reaching 50 degrees Fahrenheit. It has been assumed that as man evolved from a primitive state to an increasingly intelligent one, he developed clothing to enable him to survive in cold climates, and this enabled him to migrate from warmer zones.

We have already pointed out the impossibility of man's evolving in glaciated areas. The traditional anthropologist gets around this by saying that man evolved in warmer climates, where no special protection was necessary, and afterward migrated to colder areas when he had developed sufficient methods of protection. We have also pointed out that there is no evidence of man's evolving in what are now the warmer zones to the present temperate zones. Further, if man had evolved in a warm climate where there was plenty of game, fruit, and vegetable foods, it is extremely unlikely that he would migrate of his own choice to far more unpleasant conditions, no matter what degree of protection he had developed. And the more intelligent he had become, the less likely such a move would be.

Neanderthal man, anthropologists have discovered, lived at the same time as the earliest *Homo sapiens. The myth that Neanderthal man was the ancestor of* sapiens *has now been exploded.* In any case, the picture we have of Neanderthal man may be very far from accurate and is in many ways contradictory.

Scattered and fragmentary remains of Neanderthal man have been discovered in Europe, the Middle East, Gibraltar, and North Africa. Few anthropologists are agreed on what Neanderthal looked like. The skulls found show, in general, pronounced supraorbital ridges and a prognathous jaw, but the cranial capacity was high—1,300 to 1,500 cubic centimeters, compared to the European *Homo sapiens* average of 1,100 to 1,700 cubic centimeters. (Frequently, a smaller human brain displays a higher degree of intelligence than a larger one, so brain size in itself is no real guide to intelligence.)

Reconstructions of Neanderthal usually show him to be hairy, short, tremendously broad and muscular, with a brutish flat face, almost no neck, a hunched-over appearance, long arms, and short, bowed legs. But these reconstructions have often been subject to errors in recomposing the skeleton, or, more frequently, to the presence of bone malformations which have been ascribed to severe arthritis. (Neanderthal probably suffered from arthritis, living as he did in caves under extremely hostile climatic conditions.)

Most traces of Neanderthal man have been found in caves, usually in burial sites. Several finds have associated Neanderthal man with *Homo sapiens.* We do not know whether the Neanderthal was eliminated by *Homo sapiens,* or whether he was one of the races of *Homo sapiens;* at any rate, if we do class the Neanderthal as a separate species, he has long since become extinct. No surviving type of human being displays the characteristics associated with *Homo neanderthalensis.*

Evidence indicates that Neanderthal man was not nearly so primitive as has been supposed. Neanderthal probably used fire; round balls found in association with Neanderthal remains were used, it has been thought, as bolas to catch animals, or, heated, they could have been placed in water to heat it. They may have served both of these purposes and others we are not aware of. The fact that Neanderthal man may have used stones heated in a fire to heat water means that he must also have had containers of some kind—cooking pots, for example. It seems he may have cooked his food, perhaps by several methods, including boiling.

Graves have been found in which the dead have been carefully buried with a degree of ceremony; analysis of pollen from a Neanderthal grave in Iraq at Shanidar Cave indicates that the corpse had been buried on a bier of pine branches strewn with wildflowers. This would seem to be the action of a human being, not an animal.

It has been assumed that Neanderthal man lived for several hundred thousand years and disappeared through extinction or absorption into the later *sapiens* stock some 40,000 years ago, with the rise of the Cro-Magnon type of the *sapiens* species. But this is only guesswork.

If we assume that Neanderthal is not a species of ape-man, but a race of *Homo sapiens,* and that he lived at the same time as the rest of humankind, then it may also be argued that his age range is similar to that of *Homo sapiens.* Perhaps the Neanderthal skeletons we have found in graves do not date from the earlier appearance of *sapiens,* but from a much later period, perhaps as little as, or less than, 10,000 to 12,000 years ago.

Furthermore, Neanderthal, like *Homo sapiens,* appears to have hunted the woolly rhinoceros, the cave bear, and the mammoth. This additional piece of common ground also militates against the greater antiquity of Neanderthal.

If Neanderthal is shown now not to have been as unintelligent as was thought, why should he have chosen such an extremely unfavorable environment? The same question might be posed of *Homo sa-*

piens. Both groups are said to have inhabited areas in the northern hemisphere. The supposedly vastly increased size of the northern ice cap during their habitation of these areas would mean ice sheets stretching as far as southern England and the Mississippi Delta in North America, with ice masses stretching across most of what is now densely populated Western Europe. It is difficult, if not impossible, to understand why both Neanderthal and *Homo sapiens* should have inhabited such areas—if we assume that the polar ice cap extended so far south, or even existed at all.

Neanderthal poses a difficult problem. On the newer evidence, we can present a case that he was not a lower but a *higher* type of human being, who suffered a degree of degeneration—a view totally opposite to that formerly held.

Remarkably, the older human skeletons that have been discovered—those of Cro-Magnon—also show evidence of being superior to modern *Homo sapiens,* both in body size and in cranial capacity. Again, this seems to have been further degeneration between Cro-Magnon and modern *Homo sapiens.*

The earliest *Homo sapiens* have been called Cro-Magnon, after the district in France in which the skeletons were found. Cro-Magnon skeletons show us a very tall type of man (in excess of six feet in height), with typically European skull formation, and a cranial capacity larger than that of present-day humans.

Anthropologists have said that it may appear that *the trend since Cro-Magnon has been a decrease in brain size, and this trend may still be in operation today.* If Cro-Magnon (also possibly Neanderthal) had a bigger brain, it would seem that they may have been more intelligent than present-day human beings.

We do know that modern man of the far past had a highly developed art form. The cave paintings of animals found in many parts of Europe are clearly and accurately drawn and colored. From these drawings we have an excellent idea of the exact appearance of reindeer, woolly rhinoceros, mammoth, giant elk—animals that in some cases are now extinct. We have also found a few artifacts, some of stone, some of horn and bone and ivory, which display a degree of workmanship in no way inferior—and sometimes superior—to the products of present-day "primitive" communities.

Some rock drawings from an early period show human forms (although these are rare), *many of them apparently clothed.* More tangible evidence of this was found in a grave discovered in Russia in the regions of permafrost, which had preserved the corpse. The body was wearing fur trousers, an embroidered shirt, and necklaces and ornaments of bone and shell. Anthropologists suggest that this burial occurred some 33,000 years ago. This suggested date is very near to the figure given for the *first appearance of mankind!*

Although we have no direct evidence of a high culture for the period described as the Stone Age, there are many peculiarities which cannot be explained if we assume a primitive, emergent society. Rock paintings show people clothed, and include women in skirts, men in trousers (occasionally shorts); and they seem to be wearing shoes or boots of some kind.

Where the faces of men are shown, they appear to be clean-shaven and with the hair cut short. How did they shave and cut hair, in an age supposedly before the use of any metals? It is difficult to imagine people who lived under such unpleasant conditions as have been envisaged going to all the bother of shaving. Many ages after this, beards were the rule rather than the exception among most of the white, hairier races, possibly because of the difficulty of maintaining a clean-shaven appearance with the tools available.

Neolithic and Mesolithic man, to judge from the paintings, wore tailored clothing and the males were clean-shaven, at a date tentatively set at between 7000 to 12,000 B.C.!

Apart from the animal and human figure drawings, this prehistoric art depicts abstract patterns, some showing what are called tectiform shapes, some of which look like horses and others that could be a form of writing.

Discoveries of fine needles and buttons made of bone and ivory, some patterned by engraving, support the supposition that the wall paintings were not products of the artist's fantasy, but were patterned on reality. Such findings indicate a degree of technical attainment in the art of clothing in many ways superior to that of later ages. Buttons, for example, were unknown in Classical times, and were not used for many centuries among European cultures.

Many Neolithic skulls have been found to have been trepanned. Trepanning today is an operation in which a section of bone in the skull is removed, either to ease pressure caused by a tumor or blood clot, or to remove splinters of bone caused by a skull fracture, and the cavity closed by a plate. The operation is hardly minor, and requires great skill and care to perform. It is difficult to believe that Neolithic man—if he was, as has been thought, extremely primitive—could have carried out such operations with the crudest techniques, a flint knife, and no anesthetics or notions of hygiene.

From the growth of bone subsequent to these prehistoric operations, however, it is apparent that many of the patients survived and lived for years afterward. In Europe, until anesthetics were introduced less than two hundred years ago, operations for such things as the removal of a limb usually led to death from shock or sepsis. And operations of this type are not performed by present-day Neolithic-stage cultures.

The number of trepanned skulls is high in comparison to the total number of skeletons found, much higher than comparative numbers in present-day society. One is tempted to wonder *why* these operations were undertaken: might there have been a much more serious reason than the ones usually put forward?

Apparently early man possessed artifacts and techniques that are almost modern. Tailored clothes, smooth-shaven faces and short hair, trepanning, buttons—are these all clues to the past which we have looked at from the wrong viewpoint all these years? *In each case, instead of a gradual evolution to a higher form of civilization, we observe a regression.*

Another peculiarity, especially of Paleolithic cave art, is the frequency with which these paintings have been found in extremely dark, virtually inaccessible places. At Cabrerets (Lot), a cave system in France, narrow passages must be negotiated to reach the painted chamber. The bison of Marsoulas were painted on the walls of a narrow corridor.

Experiments have demonstrated that it is extremely difficult to mark rock walls with the flints the Paleolithic artists were presumed to have used, yet the engraved figures seem to have been executed with flowing ease. Also, these paintings, many of them in extremely confined spaces, are so precisely and accurately drawn that there must have been a good, steady illumination. Pine torches, or flickering wicks in oil-filled shells, would have given off a great deal of smoke and fumes, which, in confined and cramped conditions, would have made it virtually impossible to work for more than a few minutes at a time.

The ancient Egyptians executed highly colored painted surfaces in deep artificial chambers, like the rock tombs at Abu Simbel. Archeologists have expressed surprise at the abilities of the artists under these conditions, noting that the atmosphere would have become polluted rapidly by the kinds of lamps traditionally thought to have been used by the Egyptians.

How is it possible to assume that these ancient people used only the crudest tools and techniques and primitive lighting methods? Duplication of such equipment does not achieve anything resembling the same results, and in many cases has shown that it is impossible to duplicate these results by these methods. Unless the ancients were both stronger and cleverer than we are—in which case they could not possibly be primitives—they, too, must have found such things impossible.

One explanation of how men managed to paint under such cramped conditions, working slowly for long hours (if indeed they did), is that the Paleolithic painters belonged to a race of men who possessed *eidetic memory,* or complete recall, and were thus able to memorize every detail of what they had seen and translate this into a painting.

If this explanation is valid, then they were far superior to us intellectually. Our almost total reliance on forms of recorded information exists partly because of our imperfect memories. If ancient man possessed this faculty and applied it to painting, then it equally applied to every other undertaking. Accordingly, he should have also developed a very sophisticated civilization with remarkable speed. Yet we have almost no indication that such a high culture existed at such a remote period.

If ancient man did possess eidetic memory, and if that faculty was lost over a period of time, could there be a connection between this process and the comparatively great number of trepanned skulls?

We have a paradox: *we are told that Paleolithic man, characterized by Cro-Magnon, was physically superior to ourselves, had a larger brain capacity, and apparently also had eidetic memory—yet we find him living in the most primitive conditions.*

Several ideas may help us resolve the problem:

First, what we have discovered about prehistoric man in Neolithic times and earlier reveals the existence of just a few people. There are no towns, no villages, or even any large settlements. The physical remains of Paleolithic man discovered so far in the whole of Britain and Europe would not fill a small village.

We cannot, therefore, state with any certainty that these remains represent a static, slowly expanding human occupation, living at this time in caves and rock shelters.

Many of the rock drawings have the look of quick sketches. One is particularly struck by a group of outline drawings of the human figures in various poses, which might have come from the sketch book of any modern artist. *Were these artists*—and other people—*representative of a high civilization which for some reason was forced to lead a more primitive life-style?*

If this was so, the very extended dates assigned to these people—20,000 and 30,000 years ago—may be gross exaggerations, arising from anthropologists' suggestions that they drew and hunted species which have been extinct for some thousands of years. But is it not also possible that these animals were walking the earth only a matter of several thousand years ago? As the date of the termination of the alleged Ice Age has now been reduced to some 6,000 to 8,000 years ago, it is possible these creatures were alive up to this time. (It has recently been suggested that the American elephant was alive into historic times, and has thus only recently died out entirely.)

We can only date with any accuracy at all within the last 2,000 years or so. Even ancient Egypt's early history cannot be definitely dated. Our dates are widely at variance with the traditional historical accounts by the ancient Egyptian historian Manetho, and the writer Ipuwer. Only recently, an error in chronology of some six hundred years in biblical historical dating was found—and this from a fairly well documented era. To ascribe dates to a period where no kind of recorded history has survived must be guesswork, subject to constant revision.

We know that Paleolithic man was much like us in appearance, so much so that he would hardly be out of place in present society, and, further, that he wore surprisingly modern clothing.

There are other, more intangible things that we do not know. Paleolithic man was highly developed artistically insofar as painting and drawing are concerned. How advanced was he in music, story-telling, mathematical abilities, or language?

We can be sure of one thing: he did not use the series of grunts and basic language given in caricatures of Stone Age life. So-called primitive peoples—the Amazon Indians, remote African tribes, and the Australian aborigines—have extremely complex languages, difficult to learn and capable of as great a degree of expression as languages used in highly developed societies. These languages may stem more directly from Paleolithic or Neolithic speech patterns than from any modern language.

If we can be guided by the evidence of present-day primitive peoples, whose potential intelligence is equal to more advanced societies, and whose memory in general is more efficient (they are usually capable of learning the language of an industrialized society with greater facility than a member of that society is capable of learning their tongue), we can reason that ancient man may have been the same.

If Paleolithic man had a well-developed spoken language, with the evidence of a considerable interchange of ideas and techniques, is it not possible that he had some form of written language as well? We do not know, yet we cannot be sure that he did not. On cave walls in the Sierra Morena in southern Spain, a group of symbols was found whose age has been estimated at 20,000 years, perhaps less. Symbols of great similarity and of similar antiquity have been found carved on rock faces as far apart as Brazil and Iceland.

If the Paleolithic peoples carved meaningful symbols on rock to transmit information, it is also possible that they had more portable forms of transmitting these symbols. Neither the Indians of North America nor their cousins from Central America and Mexico were illiterate; they transmitted messages on birch bark or strips of cloth. Paper in Central America was in use from an unknown but presumably very early date. Paper, bark, and cloth would not have survived from the Paleolithic to the present, so we may never know for sure, but we can reasonably infer their existence.

So far we have discovered early traces of *Homo sapiens* that lead to some surprising inferences and even greater puzzles:

• If *Homo sapiens* had capabilities in the beginning equal to the present, is the picture which has always been presented of "Stone Age" man accurate?

• *If man was as intelligent in the beginning as he is now, and has been here some 35,000 years, why do traces of advanced urban civilization appear only during the past 6,000 years, according to our reckoning?* Why is there a gap of some 30,000 years between man's first supposed appearance and the sudden upsurge of city-building, mathematics, agriculture, medicine, irrigation?

No traditional explanation has ever satisfactorily bridged this gap. The explanation offered by Von Däniken and others—that man was shown the elements of culture by visitors from a Superior Community in space, visitors who have passed into our legends as the gods—has much to commend it.

At the same time, this answer by no means resolves all the questions. In particular, it cannot solve the problems raised by much of man's mythology, especially those legends that refer to a Golden Age, or Age of the Gods, before the Flood, and to areas of knowledge as great as, or perhaps greater than, that which we possess today. Much of this mythology appears to owe nothing to extraterrestrial influence.

Many legends *do* refer to gods who came down from the sky, but there are also extant legends about terrestrial man's flying abilities. Was this mastery of flight limited to atmospheric craft, or did ancient man—particularly if he was originally of extraterrestrial origin—retain his space-traveling knowledge?

Did there exist at some period in the past a high level of civilization, possibly worldwide in extent, even more advanced than ours today?

Was this civilization destroyed so completely that scarcely a trace remains today? Were there survivors who had to fight their way back to civilization?

We have few traces of such a high civilization. But there are certain structures for which satisfactory explanations have never been offered —structures that evidence a high degree of technical and scientific attainment.

Because no concrete trace of such a high culture exists today, we shall have to assume that it lies buried beneath present land areas, the sea, or the ice masses of the northern or southern polar regions. Remarkable discoveries have already been made, in the arctic permafrost regions of Alaska, of pre-Eskimo urban centers of considerable size. Urban cultures have not existed for thousands of years in these areas, and would not have been considered possible until they were discovered.

The main centers of a highly advanced civilization may, of course, have been completely destroyed in a vast calamity, leaving no concrete traces.

In this connection, a word about prehistoric populations is of interest. The population of the Upper Paleolithic period has been estimated at about 3,340,000, and for the Mesolithic, about 5,320,000.

Yet these figures of a small population in the Stone Age do not tell the whole story. There are legends of populations running into many millions which were destroyed almost overnight. *If these legends have any basis in reality—if destruction of life occurred on a global scale within a short period of time—then there must have been a catastrophe of tremendous proportions.* This catastrophe—the memory of it—may have been preserved in the legend of the Flood.

Ice Age versus catastrophism is one of the most controversial issues in the present scientific world, and the evidence for either or both appears interrelated. On balance, the evidence seems to show there was an Ice Age, or there was a catastrophe, but not both.

8

ICE AGE
OR CATASTROPHE?

The concept of the Ice Age is very recent. Advanced in the nineteenth century, the same century which saw the rise of the theory of evolution, it appeared to its proponents to explain certain geological peculiarities for which no alternative explanation presented itself. Curiously, however, *the Ice Age does not exist in the traditions, myths, and legends of the human race.*

According to the traditional scientific account, *Homo sapiens* appeared during the latter phase of the Pleistocene, the era of the great glaciation. His most prominent traces have been found in areas most severely affected by the glaciations. The Ice Age during which these traces were laid down ended some 10,000 years ago or less, and was replaced in Europe by climatic conditions warmer than the present. This period, known as the Climatic Optimum, lasted until some 4,000 to 5,000 years ago. Since then the climate has been steadily deteriorating.

The Pleistocene, it has been assumed, was characterized by the enlargement of the polar ice cap in the lower latitudes of the northern hemisphere. There were, according to science, four glacial periods during the Pleistocene, separated by warmer interglacial periods when the ice retreated, or partially retreated, during the past million years. The glacial periods in Europe are known as the Gunz, Mindel, Riss, and Wurm. Whether there were actually four glaciations or whether they are all aspects of the same event is a point we shall have to discuss.

As has been noted, the theory of Ice Ages was postulated principally to explain certain geological peculiarities such as the existence of moraines—great quantities of broken rock and shale—and "erratic boulders," that is, huge rocks apparently foreign to the area in which they are found. Scientists theorized that ice sheets (glaciers) moved down from the pole across the country and carried the masses of stone and great boulders with them. These solid objects were deposited when the ice melted. The number of glacial periods and their limits of advance were calculated by measuring where the various deposits and boulders lay.

If we look at the last Ice Age, the Wurm, we see the extent of this alleged ice field. At its maximum extent, it covered the British Isles as far south as the Thames Valley, North America as far south as the Mississippi Delta, and in mainland Europe, parts of Scandinavia, France, and Germany, and parts of Russia. However, it missed Jutland (Denmark), and spread no farther west over continental Europe than Mecklenburg in East Prussia. A large part of Siberia escaped.

If this *was* indeed the path of the glaciers in the northern hemisphere, called the drift (drift ice), several peculiarities remain unexplained:

• Why did the ice appear to spread in such an odd, patchy manner?

• Why is there no evidence of a similar spread of ice at the South Pole? By the same token, why are these geological conditions, which have been associated with glaciations in the past, totally absent in the present arctic regions?

• Why does the drift, which apparently carried masses of stone and clay as well as huge boulders, show no trace of having carried vegetation as well? Surely, if the glaciers were advancing into a warmer region, they would have carried *all* before them, including tree trunks and other vegetation embedded in the frozen soil. Even today, there should be *some* traces.

Another rationale for the ice-sheet hypothesis was the presence of striations, or scratches, on rock surfaces over which the ice moved. Strangely, in the Highlands of Scotland, where the evidence for the "glaciation" is more noticeable than elsewhere in the British Isles, the striations exist on the north-facing slopes but not on the south-facing slopes. If the ice, moving down from the north, ascended the hill slopes causing these scratches, they should have been even deeper on the descent of the southern side. Yet they are missing altogether on the downward slopes.

The causes of the alleged glaciations are totally unknown and no agreement has been reached among scientists regarding the theories put forward. The most popular theory is that there could be *fluctuation in the heat output from the sun,* which would make it a long-term variable. But if this were the case, there should be traces of Ice Ages at regular intervals all through earth's history. Not only are there no traces of continual Ice Ages, but there is no evidence that the sun is a long-term variable. We have already seen that the sun is a G-type star on the main sequence, characterized by great stability over hundreds of millions of years.

An alternative theory holds that the solar system may have passed through patches of extensive cosmic dust which had the effect of *obscuring and lessening the radiation from the sun.* However, there are no signs of dust clouds of sufficient density in the area of space through which the earth may have passed some ten thousand years ago.

Perhaps, other scientists say, *mountain-building* may have been responsible. According to this theory, as mountain systems like the Alps, Himalayas, Rockies, and Andes rose higher, their upper slopes would have been encased in ice. This would not only have lowered the surrounding air temperature; it would also have reflected solar radiation back into space.

On the face of it, this seems a somewhat weak theory. Considerable portions of the upper slopes of these mountain ranges are permanently covered with ice; but if they can be said to "cause" any glacial area, that area is severely limited. In the Himalayas, for example, the lower slopes and lowlands in the region maintain warm and even tropical climates. Furthermore, it has been estimated that the tectonic upheavals which created the present high mountain ranges took place in the early Tertiary subepoch. It seems unlikely that the Ice Age would have followed such a long time later.

It has also been thought that *atmospheric variations*, though so slight that they did not disturb the life of the planet, may have lowered the temperature. According to this theory, an increase in atmospheric carbon dioxide would have created a greenhouse effect by trapping more of the radiant heat of the sun. A dense, planetwide covering of vegetation may have reduced the carbon-dioxide content, which eventually lowered temperatures, which caused the Ice Age, which led to a reduction of the amount of vegetation, which eventually led to an increase in the carbon-dioxide content, resulting in an eventual rise in temperature. In this way we have a continual seesaw of temperatures over a very long period of time.

Although the theory of variations in atmospheric composition seems logical, it suffers from the same disadvantage as the variable-sun hypothesis, in that a continual succession of Ice Ages should follow all through the planet's history.

An excessive degree of *vulcanism*, another theory contends, may have in some way been responsible for the reduction of temperature, because of vast clouds of dust and ash that obscured the sun's radiation. This has probably been based on the experience of Krakatoa; the island erupted and poured vast quantities of dust and ash into the atmosphere, lowering the planetary temperature and causing a degree of weather upset for several years. Since it appears that there have been no long sustained chains of active volcanoes since Mesozoic times, this would seem to be an unlikely answer.

One further point should be raised regarding the phenomena of the Pleistocene period, especially its apparent fluctuations in temperature. Many rock masses show, in addition to striations, evidence of vitrification. Vitrification by intense heat has been observed in Scotland, on what seem to be prehistoric megalithic *artificial* structures. Similar vitrifications have been observed in the Andes in Peru, North Africa, the Middle East, and the Gobi Desert.

According to adherents of the Ice Age theory, the termination of the Wurm glaciation was followed by a period known as the Climatic Optimum, when the climate was everywhere warmer than it is now. No explanation of any sort, not even an absurd one, has ever even been offered for this development.

The Ice Age has passed from being a theory, a supposition, into the dogma of scientific fact. But *it is no more a fact than the evolution of man from some apelike creature.*

There is an alternative: the theory of *catastrophism*, which arouses controversy and bitter scorn from the defenders of the Ice Age hypothesis, apparently because it is associated in most people's mind with religion and religious myths and is therefore thoroughly unscientific.

There is no need to associate catastrophism with the wrath of God. The Flood was not necessarily a divine visitation for the sins of man. The alternative we suggest to the Ice Age theory is a series of events which had such dire consequences for what was left of the human race that later ages *attributed* these events to the anger of the gods.

Several books, notably I. Velikovsky's *Worlds in Collision* and C. Beaumont's *The Riddle of Prehistoric Britain,* have been written about the concept of catastrophism as opposed to the Ice Age. However, to the best of my knowledge, Beaumont is the only author who has come out flatly against the Ice Age theory; Velikovsky suggests that the polar regions shifted their locations.

Both of these books have as their theme a planetwide catastrophe engendered by the collision with earth of a large comet. Velikovsky dates this event in the third millennium B.C., connecting it with the biblical Exodus; Beaumont places the event in the middle of the second millennium B.C.

Velikovsky theorizes that the catastrophe was caused by a collision with a comet which is now the planet Venus, and by the close appearance of Mars when the comet (now Venus) disturbed the orbits of the inner planets. This view presupposes that the coma (head) of the comet caused a tilt in earth's axis, which in turn caused a shift in the location of the north polar ice fields, previously located farther south. The gravitational attraction caused enormous tidal forces to hurl whole ocean masses over the land surfaces. There were also side effects such as great earthquakes, the activation of many volcanoes, and the unleashing of tremendous winds caused by the disturbance of earth's atmosphere.

To a great extent, Velikovsky's theory is based on worldwide legends of catastrophic events which generally follow the sequence of conflagration, deluge, and hurricane-force winds. Darkness and rains of stones and fire mentioned in the legends are attributed to the tail of the comet as it swept through earth's atmosphere. Velikovsky amasses a great deal of carefully documented evidence to support his theory, but it has many weaknesses.

He dates the catastrophe in two parts, one 3,500 years ago, and the second some 2,600 years ago. According to *Worlds in Collision*, therefore, the final displacement of the Ice Age masses, 2,600 years ago, occurred well within documented historic times. The evidence, both legendary and geological, places the termination of the "glaciations" some 6,000 to 8,000 years ago. Furthermore, a catastrophe of the scale envisaged would have destroyed such edifices as the pyramids, Stonehenge, and other structures, which were already in existence at this time.

The main objection, however, is to the basic premise itself: that the catastrophe was caused by the action of a comet. Both Velikovsky and Beaumont credit comets with far more dangerous properties than they actually possess.

In the nineteen-forties, when their books were published, not as much was known about comets as more recent astrophysical research has revealed. We now know the composition of cometary bodies, although we are not certain how they originate or by what processes they are formed. They are thought to be members of the solar family, circling the sun in enormously elongated orbits, some of which take hundreds or even thousands of years to travel.

The size of a comet can be prodigious. Holmes' comet of 1892 had a coma (head) diameter of 1.4 million miles, and the tail of Halley's comet has been measured at 94 million miles, or more than the distance between earth and the sun. However, the size of a comet is not constant; it is usually largest when nearest the sun, shrinking as it travels away from the sun.

The coma seems to consist mainly of a great quantity of micrometeorites, some ice, frozen gases such as methane, ammonia, and carbon dioxide and molecular hydrocarbons, concentrated in the nucleus, held together loosely by gravitational attraction. However, the molecules of the gases and the micrometeorites are so widely dispersed in the body of the coma that by terrestrial standards they constitute a virtual vacuum.

The tail has a very similar composition, and spectrum analysis of the coma and tail indicate that they are not similar to planetary molecules. They consist of carbon, cyanogen, nitrogen hydride, methylidyne and hydroxyl. The more common molecules of carbon monoxide and nitrogen are ionized; that is, they have lost one electron. These are called free radicals, and exist in conditions where the atoms are too widely dispersed to capture free electrons (vacuum conditions). In this way they resemble ionized gas clouds, which exist in interstellar space and are also extremely tenuous.

When a comet is far from the sun, it is actually a loose ball of molecules of frozen gases. As it approaches the sun, both the coma and the tail grow in size. The stream of particles comprising the solar wind, and solar radiation itself, cause the change in size. Gases are given off by the warming comet. They emit their own light and are driven away from the sun by radiation (light) pressure from the sun. This is why the tail of a comet always faces away from the sun, regardless of the position of the comet in relation to the primary.

The mass of a comet, considering its size, is extremely slight. Dr. Elizabeth Roemer at the University of Arizona has estimated that the mass of the comet Wirtanen 1956C was about 1,015 pounds terrestrial. On earth this would comprise a rocky ball with a diameter of two miles, or a ball of ice and snow with a diameter of three and a half miles.

Even if a comet were a solid mass of the order described, it is unlikely that it would cause more than localized damage at the point of impact, especially since this mass would be drastically reduced by vaporization on its passage through earth's atmosphere. (As a ball of ice and snow, it would probably be completely vaporized before reaching surface level.) As comets are far from solid, it is more likely to be harmlessly dissipated in the upper atmosphere.

If Venus had once been a comet, it would have to have been staggeringly large. In both size and mass, the present planet Venus is almost the twin of earth; in a state as diffuse as would be true of a comet, Venus would have to be at least as large as the sun itself. This is a technical absurdity; a comet would not form in so large a diffuse mass. Furthermore, its size, even millions of miles distant from earth, would be so great that its glowing mass would fill the entire sky. No phenomenon of this magnitude has been recorded in the annals of the human race.

Finally, if a body roughly the size of earth, namely Venus, had approached near enough (Velikovsky says their atmospheres intermingled), there would have been very little possibility of close contact. Even if they had approached close enough for atmospheric contact, the frictional stresses involved would have vaporized the atmospheres of both worlds.

However, the factor known as Roche's Limit would have made

this unlikely: if the moon were to approach earth too closely, for example, gravitational forces would shatter the smaller body—the moon—before any actual contact. An exact, literal collision between earth and the moon could not happen; the moon would shatter long before it reached the outer fringes of earth's atmosphere. Damage to earth would be immense, but the planet would survive, and would be surrounded by the moon's debris, a ring system very similar to that of Saturn. Eventually, the debris would fall to earth in showers of meteorites.

In 1913, Hans Hoerbiger advanced the now discredited theory that the catastrophe occurred when a moon of ice fell onto earth. This, he said, not only led to convulsions but deposited vast amounts of ice on earth which led to a cooling of the climate.

Since earth is slightly more massive than Venus, a similar fate would overtake Venus. It is possible that the interplay of gravitational fields between Venus and earth would have equally disastrous effects on earth, and it may well be that both planets would be shattered, in which case the debris of the two planets would form an additional asteroid belt within the solar system. Even if this did not happen, it is certain that earth would have its atmosphere vaporized or dissipated by the strong gravitational pull of Venus before Venus fragmented.

Had there been a close contact between Venus and earth in the past, Venus would no longer be in the sky, and even if earth survived, there would be no one here to notice its disappearance. Our world would have been long since devoid of both life and atmosphere.

Leaving aside the question of what caused the catastrophe for the moment, we shall examine what further evidence supports this idea, as opposed to an Ice Age.

It is now generally agreed that the alleged Ice Age came to an end some 6,000 to 10,000 years ago, but it is not agreed, even by the adherents of the Ice Age hypothesis, whether this was a gradual process or was sudden and cataclysmic. Recent discoveries have cast serious doubt on the thesis of a gradual retreat of ice masses over many thousands of years.

Most books dealing with the Ice Ages tend to ignore, or mention only briefly, the fact that entire species of animals disappeared forever at the termination of the last glaciation. When it is mentioned, we are rarely told the circumstances, which do not fit in with a gradual retreat of ice sheets over many thousands of years. *These disappearances are explicable only in terms of a great catastrophe.*

9   THE GREAT EXTINCTION

The termination of the Wurm glaciation ushered in the greatest extinction of life known in recent times. This extinction included most of mankind.

The mammoth, the largest mammal of the Pleistocene, inhabited large areas of the northern hemisphere, the warm Siberian plains, and the southern latitudes of Europe. Contemporary with the mammoth, the mastodon, or American elephant, held sway in North America, Mexico, and Central America.

There does not seem any reason—any *natural* reason—why this species ceased to survive.

The mammoth was a highly developed member of the elephant family, comparable in size to the present-day Indian elephant. Its tusks were often ten feet in length, and its teeth were highly developed, with a greater density than any other of the elephants, antique or modern. It was ideally fit for an herbivorous way of life and for survival under suitable conditions.

In 1799 bodies of mammoths were found in the Siberian tundra, in the regions of present permafrost. *The bodies were perfectly preserved* and sledge dogs ate the flesh without ill effects. The flesh was firm and marbled with fat and appeared as fresh as well-frozen beef.

In line with the theory of a gradual retreat of the Ice Age and geological events taking place over great expanses of time, it was supposed that a gradual sinking of the land forced the mammoths to the hills, where they were isolated in marshes. If this had been such a slow process, the mammoths would not have been so trapped. Being trapped this way, it was reasoned, they starved to death. But the mammoths did not die of starvation. *In their mouths and stomachs were found grass and leaves, still undigested, proof that they died suddenly and without warning.* Moreover, the vegetation the mammoths fed upon does not now grow in Siberia, but a thousand miles farther south. Apparently the Siberian climate has changed radically since the end of the alleged glaciation.

After arctic storms, tusks of mammoths have been washed ashore on arctic islands. Part of the land where they lived has since been covered by the Arctic Ocean.

Frozen mammoth corpses have been found, preserved in masses of ice, in either a standing or a kneeling position, with food in their mouths and in their stomachs. Could these mammoths have lived among the ice and snows of the Ice Age, and starved to death as the Ice Age ended slowly? Had they died slowly, either of starvation, or by drowning in marshes, they would have been found lying down. They would not have been complete animals whose stomach contents could be examined; they would have been mere skeletons. An animal standing, in the act of eating, then preserved in ice for all to see, *has been killed and instantly frozen, in a matter of minutes, where it stood.*

The mammoth was not the only animal affected. In an article in the *Saturday Evening Post* in 1960, a noted scientist wrote:

"About a seventh of the entire land surface of our earth, stretching in a great swath around the Arctic Circle, is permanently frozen ... the greater part of it is covered with a layer, varying in thickness from a few feet to more than a thousand feet, composed of different substances. It includes a high proportion of earth or loam, and often also masses of bones or even whole animals in various stages of preservation or decomposition.

"The list of animals thawed out of this mess would cover several pages ... the greatest riddle, however, is when, why and how did all these creatures, and in such absolutely countless numbers, get killed, mashed up and frozen into this horrific indecency?

"These animal remains were not in deltas, swamps or estuaries, but were scattered all over the country. Many of these animals were perfectly fresh, whole and undamaged, and still either standing, or at least kneeling upright.

"Vast herds of enormous, well-fed beasts, beasts not specifically designed for extreme cold, [were apparently] placidly feeding in sunny pastures at a temperature in which we would probably not even have needed a coat. Suddenly they were all killed without any visible sign of violence and before they could so much as swallow a last mouthful of food, and then were quick-frozen so rapidly that every cell in their bodies is perfectly preserved."

Another article, by C. H. Hapgood, published in the *Saturday Evening Post* in 1959, stated:

"One of these periods of wholesale destruction of life occurred at the end of the last Ice Age. That was a natural disaster, which, according to one writer, destroyed some 40 million animals in North America alone. In a few thousand years life on earth assumed a radically new aspect. It is apparent that millions of animals once flourished in areas *now bitterly cold.*

"By the use of the radiocarbon dating method, scientists revised the date of the end of the last Ice Age, making it only 10,000 years ago instead of 30,000 years.

"The other new method of dating, which we call the ionium method, has revealed that during the last million years, Antarctica has several times been non-glacial. When these cores from the bottom of the Ross Sea were dated it was found that the most recent Ice Age in the Ross Sea *began* only 6,000 years ago."

Only a few thousand years ago, then, vast numbers of animals perished in a cataclysm. Legends from all parts of the world suggest that thousands, perhaps millions, of human beings also perished in such a cataclysm.

In addition to evidence of vast climatic changes which have transformed huge areas from warm climate to their present arctic nature, we have evidence that other areas of earth have undergone different alterations in climate. At the present time, desert areas cover much of earth: North Africa and the Sahara, most of the Middle East, the Gobi in Asia, the Kalahari in Africa, the arid regions of the

American Southwest, the Peruvian coastal desert of South America, and most of the interior of Australia.

There is a great deal of evidence that this was not always so:

• Rock paintings in the southern regions of the Sahara show a great number of species of animals—antelope, giraffe, and others—which now live much farther south.

• Paintings and artifacts in the ancient urban center of Catal Huyuk in Anatolian Turkey show that the now desolate plains below the Taurus Mountains were once grassy savannahs occupied by huge herds of ungulates.

• The description in the Old Testament of the Middle East lands as "flowing with milk and honey" could not possibly refer to recent historical times.

• The now arid regions of the coastal strip of Peru and Bolivia must once have been very different. Deserts could not have supported the extensive cities with great urban populations whose ruins have been uncovered there.

• Mayan legends describe the Yucatán as the land of the "honey and the deer," yet much of the interior of the Yucatán today is uninhabited and uninhabitable.

• A little over two thousand years ago, North Africa was the granary of Europe, a well-watered, fertile land bordering on the Mediterranean. Vast wheat fields and dozens of Roman towns and cities lay in this region. The ruins of these cities lie buried under shifting desert sands today.

• The Gobi Desert also exhibits traces of once flourishing flora and fauna, all of which have now vanished.

Separately, these facts do not seem to be of any great consequence. But if we try to make a coherent picture of them all, a most extraordinary and significant fact emerges:

Before the catastrophe, both the northern and southern polar regions were warm, and the deserts were nonexistent. Earth, it appears, was warm, well watered, green, and fruitful. Both the Pleistocene and the Holocene (the present), we are also told, are abnormal periods in earth's climate.

We cannot be sure that there were other glacial periods in earth's history, but we do know that *we are now in a cold phase*, that is, that large areas of earth are glaciated. The only concession most geologists will make to the hypothesis that catastrophic climatic changes occurred is that the glaciated zones may have changed their positions due to sudden shifts of earth's axis, an idea that most astronomers dismiss.

It is also possible that the deterioration in climate which happened some 6,000 to 10,000 years ago is still in progress. During the past 4,000 years, the desert areas have increased their size. The areas which nurtured the Sumerian, Assyrian, and Babylonian civilizations, and which were then fertile, have become so arid that few people now live where the ruins of once great cities now lie buried.

It has been thought that the ice mass at the South Pole is becoming greater. H. A. Brown, an engineer from the University of California, maintains that the antarctic ice cap, presently 8,200 feet thick at its maximum and comprising 7.2 million cubic miles of ice, is still growing. Eventually, he says, the equatorial bulge will be unable to maintain the equilibrium of earth against this growing mass of ice, and earth will tilt or oscillate to find a new center of equilibrium. This will lead to catastrophic conditions, which, he suggests, has happened before.

More puzzling evidence suggests catastrophic changes:

• Remnants of palm, fig, and magnolia trees have been found in arctic lands, and coral reefs once flourished in Spitzbergen.

• The presence of coal in Antarctica shows that the region was once covered in forests. It is not enough to hypothesize a warmer climate in this region. Vegetation of this kind could hardly have survived under the present conditions of six months day and six months night.

The axial tilt of the earth may at one time have been different from what it is today.

10   ARE WE LIVING IN THE ICE AGE?

Prior to the period that began 6,000 to 10,000 years ago, the polar regions were warm and filled with plant and animal life. The deserts were green plains, nurturing vast herds of animals. All of this, *well within the lifetime of the human race,* contrasts sharply with the traditional picture of man living amid the barren wastes of the Ice Age, making a living by hunting polar beasts.

Curiously, *the memory of the human race,* as expressed through myth, legend, and tradition, *holds with the first, not the second, picture.*

Scientists are generally agreed that for long periods of the planet's history, earth *was* warmer than it is now; it is only in the matter of the Pleistocene glaciations that they part company with advocates of the catastrophe theory.

How do we explain this phenomenon? If the sun was not warmer in the past (and astrophysical evidence is against this in the case of our sun), then the alternative is that earth was once slightly nearer the sun.

Being nearer, the sun would not only make earth warmer; it would also, by decreasing the orbital distance to be traversed, *make a shorter year.*

This would explain the puzzling difference between ancient calendars and those of more recent times. Documentary evidence from the ancient world shows that calendars described a shorter year than the present, and that due to the "changed order of things" the calendars had to be adjusted to account for an additional five and a quarter days:

• The Reverend Bowles, a nineteenth-century archeologist and authority on megalithic monuments in Britain, says that the circles of Avebury represent a calendar of 360 days, and that an extra five days were added later.

• In all the ancient classic writings of the Hindu Aryans, there is a year of 360 days. The *Aryabhatiya,* the ancient Indian mathematical and astronomical work, says: "A year consists of 12 months. A month consists of 30 days. A day consists of 60 nadis. A nadi consists of 60 vinadikas."

• The ancient Babylonian year was of 12 months of 30 days each. The Babylonian zodiac was divided into 36 decans, this being the space the sun covered in relation to the fixed stars during a 10-day period. Thus the 36 decans require a year of only 360 days. Ctesias wrote that the walls of Babylon were 360 furlongs in circumference, "as many as there are days in the year."

• The Egyptian year was originally 12 months of 30 days each, according to the Ebers Papyrus. A tablet discovered at Tanis in the Nile Delta in 1866 reveals that in the ninth year of Ptolemy Euergetes (ca. 237 B.C.), the priests at Canopus decreed that it was "necessary to harmonize the calendar according to the present arrangement of the world." One day was ordered to be added every four years to the 360 days, and to the five days which were afterward ordered to be added.

• The ancient Romans also had a year of 360 days. Plutarch, in his life of Numa, wrote that in the time of Romulus the year was made up of twelve 30-day months.

• The Mayan year was of 360 days, called a tun. Five days were later added, and an extra day every fourth year. The Mayans computed the synodal period of the moon as 29.5209 days, as accurately as we can calculate today with our sophisticated equipment. Their degree of accuracy would surely not have been less when they computed the 360-day year. "They did reckon them apart, and called them the days of nothing; during which the people did not anything," wrote J. de Acosta, an early writer on America.

• The Mexicans at the time of the Spanish Conquest called each 30-day period a moon.

• The Incan year was divided into 12 quilla, or moons of 30 days. Five days were added at the end, and an extra day every 4 years. The extra days were regarded as unlucky, or fateful.

• The ancient Chinese calendar was a 12-month year of 30 days each. They added 5¼ days to the year, and also divided the sphere into 365¼ degrees, adopting the new length of the year into geometry as well.

This extraordinary state of affairs has been explained by scholars' saying that the ancients first proposed a rough system of yearly count, and afterward refined it as their mathematical knowledge increased. Their explanation seems illogical; the peoples in question were already excellent mathematicians and astronomers when they created the 360-day calendar. Why should they have made such mistakes and then rectified them later? Furthermore, *one group or culture could have made such a mistake, but would so many in such widely scattered regions of the world make the same mistake, and then rectify it in the same way?*

The ancients' subsequent addition of 5 days to the calendar, and an extra day every 4 years, indicated that they knew that the year was composed of 365¼ days, which we now know to be mathematically precise. As this superseded the previous year of 360 days, they must also have computed the length of that year. If their mathematics were exact in the one, there is no reason to suppose that they were less exact in the other. Furthermore, the reason given for the recalculated calendars was not that they were in error, or that they had improved on more primitive techniques, but that there was a "changed order of things."

An alteration in earth's distance from the sun and its orbital position would account for the difference in the length of the year. If earth had been jolted out of its previously held orbital position, the moon also would have been affected. The change would have affected the earth/moon system. Since the moon is a smaller body than earth, and the distances between them much smaller than between the earth and the sun, the differences would have been even more noticeable in the case of the earth/moon system than in the case of the earth/sun system.

This would appear to have been the case. In several ancient sources it has been found that there were 4 9-day weeks to each lunar month, making a month of 36 days. This 9-day phase has been found in ancient Greek, Babylonian, Chinese, and Roman sources, among others. As these lunar computations did not fit with a year of 360 days, the calendars were altered to a 10-month year. This was an attempt to regulate the "new" year to fit the "old" 360-day year.

In the Pacific, the Chams of Indochina have a calendar of 10 months, as do the inhabitants of the Gilbert Islands and the Marquesas. In every case these months are based on the lunar cycle, which, to make a 10-month year, must have been longer than the present month.

Scholars are quoted in *Worlds in Collision* regarding the records on tablets discovered at Nineveh: "How could the star-gazers who composed the earlier tablets be so careless as to maintain that the year is 360 days long, a mistake that in six years accumulates to a full month of divergence; or how could the astronomers of the royal observatory announce to the king the movements of the moon and its phases on wrong dates, though a child can tell when the moon is new, and then record all this in very scholarly tablets requiring advanced mathematical knowledge?"

From examination of calculations made in antiquity regarding the sun/earth system, it appears that the year had to be revised from a 360-day year to one of 365¼ days. It also appears that the month varied from 36 days maximum lunation to the present 29½ days (approximately).

If the earth had assumed a new orbit and axial inclination as a result of a catastrophe, with subsequent perturbations in the earth/moon system, it is probable that there would be a period of fluctuation before the earth/moon system settled into its now apparently stabilized orbit. If there is still a residual effect from the original displacement, it is by now very small, and it may be several centuries before any significant increase in the length of either the day or year may be noticed.

The rearrangement of the calendars in antiquity has long been a puzzle to scholars, and cannot lightly be dismissed, as the ancients were as good mathematicians as we ourselves are, perhaps better. *This can only be explained by a great and significant change in earth's orbital position,* with its consequent disruptive effect on the earth/moon relationship.

The hypothesis that a year before the Flood—or catastrophe—was of 360 days' duration and not the present 365¼ days is not as extraordinary as it sounds. Scientists have said that fossil shells have revealed that 500 million years ago, the length of the year was some 412 days.

Two explanations could account for the fact that the year was shorter by 5¼ days before the catastrophe. Either earth's orbit was less elliptical than it is today, or earth was slightly nearer the sun. Since the polar regions were not at this time subzero in temperature, but warm, and since the day-and-night pattern must have been different from the present one to enable vegetation to flourish in these areas, we conclude that earth's axial inclination must have been different from what it is today. Any or all of these factors would have made an enormous difference in the terrestrial climate.

Leaving aside the matter of what caused this shift, let us assume that the earth, at the time prior to the catastrophe, was orbiting the sun at a more favorable position than it does at present.

Evidence shows that the climate of the whole planet was warmer throughout the year. If earth were receiving more heat from the sun, one further effect would occur, which has also been taken as evidence of an Ice Age: the lowering of the ocean levels. It has been assumed that the absorption of large masses of water within the glaciers and ice sheets would have lowered the levels of the oceans by over 200 feet, compared with their present levels.

If earth were nearer the sun, the same effect would have occurred —by *evaporation*. Greater heat would have meant greater evaporation from all surface waters, which would have produced a warmer —and damper—planet. This would seem a much more reasonable explanation for the lack of deserts during this period than the Ice Age model. The increase in heat and subsequent higher rate of evaporation would have meant a much higher concentration of water vapor in earth's atmosphere.

Written and traditional sources substantiate this claim: for example, Genesis 1:6-7: "God said, Let there be a vault between the waters, to separate water from water. So God made the vault, and separated the water under the vault from the water above it, and so it was; and God called the vault heaven."

If we translate this Genesis statement into the modern idiom, the waters under the vault are rivers, seas, lakes, and oceans, and the waters above are the clouds and the high-altitude layers of suspended water vapor.

This concept expressed in Genesis must be of extreme antiquity. The material of the Old Testament, much of which was collected together and rewritten in the fifth to sixth centuries B.C., covers a period beginning in the twelfth century B.C. But we have no idea how old some of this Old Testament material may be. A document written in the twelfth century B.C. may be the recording of an oral tradition of an unknown antiquity. Or it may have been copied from an earlier, decaying manuscript, which in turn could have been a copy of an even earlier manuscript. We can have no idea of the original source, or the date of the original manuscript. The earliest versions could date back many thousands of years.

The early part of Genesis deals with a period prior to the Flood. The Hebrews of the Middle East in the period we call historical lived in a region already semiarid in most parts, with extensive desert areas. They would not have conceived of a water-vapor layer above earth; this must have been knowledge passed down to them from a remote period by people who *were* aware of this. By 500 B.C. these words were still being faithfully copied by the rabbis and scholars; it did not matter that they might not understand what they wrote, provided that they recorded the divine message correctly.

We also cannot be sure that the original source was the Middle East area. Although a legend may be found incorporated in the literature of a certain country, the original source may actually be far removed from that particular location. If our hypothesis about the catastrophe is correct, sources for such "myths" would probably have come from more northerly zones.

In the period before the catastrophe, we have seen that the world was a vastly different place. Many scientists agree that the land area may have been much more extensive than it is at present. The British Isles, Greenland, Iceland, and Scandinavia may have formed one land mass. On the other side of the world, the Gulf of Mexico with its present island groups may have been dry land, and the continental outlines of the Americas may have extended to include now submerged portions of the continental shelf.

In recent years, divers have claimed to see the outlines of ancient buildings both in the Gulf of Mexico and off the coast of South America in the inshore waters of the Pacific. Such sightings would seem to confirm that the land area had been greater than today. Most traces of ancient man may lie in these areas, now sunk beneath the sea.

We do not know the distribution of population on the earth in the pre-Flood period, but the figures given by anthropologists for a period before 10,000 years ago—a little over 5 million—could be wildly inaccurate. These figures appear to be disputed by many ancient sources:

• An ancient Indian Aryan saga, the *Mahabharata* (about which we shall have more to say later), says that "60 million people in great cities were killed in one dreadful night" (during a great catastrophe).

• The Troano Manuscript of the Maya says, "The lands of the West (Mu) were continually shaken in the night. Twice upheaved, they broke into ten pieces and sank, together with the millions of inhabitants."

• In the biblical Genesis, although no figures are quoted, it is said that before the Flood, man spread over the earth.

Legends from many parts of the world speak of huge populations, running into many millions, and great cities. These legends must tell of an unknown period in earth's history, that which we call prehistoric, for in the historic period, cities with millions of inhabitants did not exist. The great cities of antiquity, such as Ur, or Babylon, Cuzco of the Incas, or Tenochtitlán of the Aztecs, all had populations below the million mark.

No traces of cities like those mentioned in the legends exist. If any traces of the cities that vanished during the catastrophe do still exist, they are probably buried beneath the sea, or even under the bed of the sea. The chances of our being able to discover such ruins are minute.

A glance at a modern map of the world shows that, with very few exceptions, the largest cities lie near or on the coast. The three largest and most densely populated cities—London, New York, and Tokyo —are coastal seaports, as are San Francisco, Singapore, Los Angeles, Colombo, Rangoon, Hong Kong, and Sydney. If the oceans of the world were to rise by 200 feet, most of the world's principal cities would vanish. There is no reason to suppose that the situation would have been materially different in ancient times: major cities were probably located on or near the then coastal regions.

Even if they had not been entirely destroyed in a catastrophe, there is every chance that they would not only be covered by hundreds of feet of water but buried below hundreds of feet of mud and silt. The most sophisticated underwater research equipment would have a difficult time uncovering them.

Even the ruins of cities vastly more accessible, such as those buried under the shifting sands of Mesopotamia, or hidden in mounds in Turkey or the regions of Middle America, lay buried and forgotten, in many cases for thousands of years, until exploration in the nineteenth and twentieth centuries. Many more ruins of cities lie buried in jungles in the foothills of the Andes. Only within the last two decades has it been found that at Teotihuacán in Mexico, in addition to the already known pyramid complex, there was a large residential city, with an estimated population in excess of a hundred thousand. Many of the megalithic prehistoric monuments in Britain have scarcely been charted or investigated.

Most artifacts, including those made of stone, have a limited life span; the ravages of climate, the action of water and the atmosphere, may have eradicated all trace of man's handiwork after a period of six or seven thousand years. We do not, therefore, have a great deal of concrete evidence to show that there was a highly advanced civilization prior to a great catastrophe. But there is a certain amount of concrete evidence that has been either overlooked or denied. Such evidence lends support to a theory claiming the existence of a highly cultured state at a remote period, while it can hardly be used to support such theories as the existence of an Ice Age or the evolution of man.

There is, however, much more than physical evidence to rely on. Further evidence can be found in the existence of certain fields of knowledge which do not conform to the theory that the human race emerged from pure barbarism to civilization. Knowledge can be totally lost when there is no longer any intelligent creature to transmit the information. Although, with the passage of time, a certain amount of knowledge may be lost or forgotten, a certain amount will remain, even in a distorted or garbled form.

The concept of a Golden Age, or a Paradise World, destroyed in a great disaster, usually described as the work of the gods, is so widespread and deeply rooted in human mythology that it almost appears to be a piece of instinctive knowledge in the collective human subconscious.

We have already assumed that the world was a vastly different place climatically before the catastrophe. The human race as a whole may also have been vastly different. *The fall of man may be not mythological, but an actual event in which the catastrophe ushered in the decline of the human species.*

11   DID PEOPLE BEFORE THE FLOOD KNOW MORE THAN WE DO?

Before we proceed any further, we should perhaps deal with a very thorny problem which we have only once touched upon slightly: the question of different human races.

We have already. mentioned the virtual impossibility of three different kinds of prehumans giving rise to three different kinds of human beings by chance mutation. Yet three major groups exist: white (Caucasoid), yellow (Mongoloid), and black (Negroid). All the other races, variants on these three basic stocks, safely interbreed. At some time in the past they must have had a common ancestor.

Evolution has failed to show how these three different groups may have arisen. Even the view that these three groups differ as the result of climatic adaptation does not seem to supply a complete answer. It has been suggested that the Negroid is black because of an increase of melanin under the effect of strong sunlight, which assists in the protection of the organism against excessive ultraviolet radiation.

(This is why even the white races develop a tan when exposed to a greater degree of sunlight.) However, it has been found that those South American Indian tribes of Mongoloid origin—therefore, pale skin—when kept from exposure to strong sunlight, retain the tan coloring of their Mongol ancestry. Negroid children, on the other hand, are born black. It seems that climatic adaptation is only part of the answer.

Further, it has been suggested that the flattened face and the eye fold (a layer of fat beneath the skin) of the Mongoloid is a climatic protection against the bitter winds of Asiatic winters. However, many groups of Mongoloids who have lived in tropical climates for many thousands of years are no less flat-featured or slant-eyed as a result. If adaptation to environment were at work, there should by now be some sign that at least some Mongols living in the tropics exhibit differences due to climatic adaptation. This has not happened.

The Caucasoid races, who live mainly in temperate or cold climates, show no evidence of an adaptation to such conditions. By the yardstick applied to Mongoloids for climatic adaptation, the Caucasoids should exhibit characteristics of flatter features and slanted eyes. Interestingly, the Negroid is excessively flat-featured—once again contradicting the notion that this characteristic is a "cold-climate adaptation."

By the standards applied to the Mongols and the Negroes, the Negro should be dark-colored and sharp-featured, while the European should be flat of face and have slanted eyes.

Negroes whose families have lived for many generations in the temperate zones such as the United States show no signs of becoming more European-featured or less dark-skinned, except as a result of interbreeding with other races. At the same time, they are perfectly adapted to life in a temperate zone, and experience no more discomfort during cold weather than their white compatriots.

Furthermore, while members of the white race can adapt with little difficulty to tropical conditions, white people born and bred for generations in tropical climates show no tendency to be born with darker skin. Descendants of Europeans in India from the time of Alexander the Great—over 2,000 years ago—still produce "throwback" children of pale complexions and fair hair.

We have theorized from both physical evidence and mythological sources that earth was warm from pole to pole when man entered the picture. Although this would make the theory of a natural evolution more tenable, it would also have made the emergence of three separate races adapted to differing climatic conditions both unlikely and unnecessary. Adaptation to the climate of earth as we know it does not seem to supply the answer. Yet the ability of all races to interbreed successfully implies a common origin.

We have hypothesized that man on earth may be the descendant of colonists from elsewhere in space. Is it not possible that humanoids living under different suns would develop characteristics that would assist in their survival? There seems no reason why the Negroid should have a darker skin than the white on this particular planet, under this sun; but under the radiation of a star somewhat different from our sun, this coloration may have rendered a higher degree of protection. The Mongoloid may have developed on a colder planet, under a sun emitting less radiation than ours. This thesis is no less likely than the explanations offered by evolutionary anthropologists.

In any event, supposing such was the case, the Superior Community may have decided to send three humanoid groups to settle this world to see which would most successfully adapt. All have adapted equally successfully.

We could look at this from the opposite point of view. If we undertake the colonization of an extrasolar planet, may we not do the same thing? We might very well send Indians from the high altitudes of the Andes, Negroes from the low-lying West African coast, Europeans from Britain or mainland Europe. We would then be introducing to another world three racial groups who may, in the course of time, wonder why and how three such groups came into being.

It may be argued that physically we resemble the other animal life of the planet too much to have come from anywhere but this planet. The basic design and biological processes of all mammalian forms are very similar. However, as we have already pointed out, both nuclear and biological processes follow basic laws everywhere in the universe, and carbon-based life forms will follow a similar pattern, whether it be on earth or on a hypothetical planet circling Tau Ceti. Terrestrial animals of this basic "design" appear to be highly successful from a survival point of view, and would probably follow the same successful pattern on other similar planets.

Once again, we can look at this problem from the opposite end. What would be the case if earth people settled on an earth-type planet elsewhere in the galaxy? Another earth-type world would have a similar atmospheric composition, with a similar temperature range. We can therefore expect to find vegetation, operating through photosynthesis, chlorophyll-based, and therefore green.

The animal life would be of the same order as terrestrial forms: oxygen breathers, ingesting vegetation and disposing of waste matter. Such animal life would have organs for sight, smell, hearing, and making sounds, as is natural in an atmosphere where sound waves will carry. In short, there would be many forms similar to those found on earth; differences would probably be in size, coloring, and superficial detail.

It is possible that in the course of time, our extrasolar colonists would forget their origins and generate "creation myths." They could also claim that they were truly indigenous to the planet by pointing out that they had the same basic resemblance to the other forms which existed there.

We can have no idea of the detailed plan of extraterrestrial colonization which may have been adapted. Perhaps the three racial groups initially occupied different zones of earth; perhaps they intermingled. The relatively clear definition among the three groups even today suggests either that they may have occupied different zones or that a solitary restriction forbade interbreeding. It does seem possible that in the past the whites occupied the northerly zones of earth, the Mongoloids Asia, and the Negroids Middle and South America.

> Curiously, there is little trace of human occupation of any sort in Africa south of the present Sahara until less than 10,000 years ago, which could be the period *following* the catastrophe.

It would also appear that color-consciousness and racial antagonism is of comparatively recent date, since even as late as Roman times little attention was paid to color as such. Special mention of differently colored racial minorities in the past are conspicuously absent in ancient writings. The sacred book of the Maya, the *Popol Vuh,* for example, states that "the white, yellow, and black men lived altogether in harmony, under the beneficent rule of the white men."

Physical evidence seems to corroborate the text: giant heads of a definite Negroid cast, the so-called Olmec heads, have been found in Central America, though many of the present-day descendants of the Maya would seem to be of Mongol extraction. The most ancient Maya legends say that the earliest Maya cities in Central America were built by the Saiyam Uinicob—the "Old White Fathers"—during the "Great Arrival," and carved representations have been found of bearded men with Caucasoid features.

We have no real knowledge of how long true man has been on the earth. Scientific estimates usually say some 35,000 to 40,000 years, although the figure could be as low as 20,000 years. Curiously, little attention has been paid to mythological and legendary records of man's history on earth:

• Maya calendric stelae record events going back 30,000 years. As these stelae may have been erected some 2,000 years ago, this is some 32,000 years ago.

• The Egyptians, too, gave an exact dating for the commencement of the reign of the demigods who ruled before mankind. This dating was based on two calendars. One divided the year into 365 days, taking no account of leap years. The other was an astronomically exact solar calendar, based on the rising of the dog star (Sirius) at dawn. By the two different methods, it takes 1,461 years for the calendars to coincide, and the calculations purported to show 25 such coincidental points: 25 x 1,461 years, or 36,525 years from the rule of the demigods to the thirtieth Egyptian dynasty in 332 B.C. The computations, based on the rearranged calendar in use in the first millennium B.C., add up to just over 38,000 years from the present time.

• If we add together the ages of the descendants of Adam to the time of Noah (the Flood) as recorded in Genesis, we arrive at a figure in excess of 9,000 years, which, added to the approximate date from then to the present, makes a total of about 20,000 years.

Although discrepancies between the biblical figures and those of the Maya and the Egyptians are considerable, perhaps some of the generations have been omitted or forgotten by biblical chroniclers, which may explain part of the difference. All the figures fall within the limits of the figures arrived at by our scientists for the appearance of the first true *Homo sapiens*. It would seem that *mythology has already told us what scientists have so laboriously corroborated: that* Homo sapiens *has inhabited the earth for less than 50,000 years, somewhere between 36,000 and 20,000 years.* As this figure agrees so closely with modern scientific estimates, we can reasonably assume that this mythology is not, properly speaking, mythical, but is based on facts and knowledge of human history.

From many other odd pieces of both knowledge and artifacts, we can infer the possibility, at least, that a Superior Culture existed before a great catastrophe:

In a temple in India, for example, archeologists have found a remarkable calendar that calculated time as far back as the point at which the earth had been formed by Brahma. The computations cover a time span of some 4.5 billion years (4.5 eons). Scientists estimate the age of the earth at between 4.5 and 5.5 eons. The figures are too close, surely, to be coincidental.

What is remarkable about such calculations—it is as true of Maya as of Indian mythology—is the *large numbers involved*. Both employ figures running into many millions. This is not a characteristic of primitive peoples. Even the sophisticated Romans had no symbols for figures over a thousand. It would seem that the Indian and Maya cultures—in the area of mathematics, at least—must have descended from societies which were highly advanced mathematically. As mathematics is the basis of many sciences, we can infer that this original source represents a highly advanced civilization, which had disappeared.

In the book of Genesis, great ages were attributed to the descendants of Adam before the Flood. Adam himself is reported to have lived 930 years, Seth 912 years, Enosh 905 years, Kenan 910 years. According to Sumerian mythology, too, the gods before the Flood lived for enormous periods of time. (Actually, Hebrew and Sumerian myths are so nearly identical that they could have sprung from a common source.)

Another curiosity of both Hebrew and Sumerian myths is that they record a steady *decrease* in the life span of man after the Flood, as we note from Genesis:

"Shem was a hundred years old when he begat Arphaxad, two years after the Flood. After the birth of Arphaxad, he lived five hundred years.

"Arphaxad lived for four hundred and thirty eight years.

"Peleg lived two hundred and thirty nine years.

"Nahor lived one hundred and forty eight years.

"Nahor was the grandfather of Abraham, the founder of the

Twelve Tribes, and who left Ur, his home city, in the land of Shinar [Sumer]."

Gilgamesh, the culture hero of the Sumerians, reigned after the flood for 102 years.

What is remarkable about these figures is that 1) they are precise as to the number of years; and 2) they show a progressive *shortening* of the life span with every generation after the Flood.

If we add together the ages of Noah's descendants until the time of the founding of Sumer by Nimrud, we come up with a figure of about 3,000 years. The Sumerian civilization is estimated to have begun about 3000 B.C. The combined total is between 6,000 and 7,000 years—a reasonable estimate of time elapsed between the Flood and the present. The glaciation of Antarctica—one of the consequences of the catastrophe—has now been estimated as beginning some 6,000 years ago. Again, *mythology and science seem to agree.*

Some scientists have asserted that the great ages attributed to the ancients were gross exaggerations, that these figures were ascribed to make the ancients seem "godlike" in their longevity. The exceedingly long lives of pre-Flood people play an important part in the mythology of the "Age of Gods"—*except that they were not gods, but men.*

The difference between these pre-Flood people and those of the present was that the former lived for vast spans of time, apparently free from "sin"—bodily, or mental, afflictions. We have already mentioned that a major factor of an interstellar community may be extreme longevity, and an advanced civilization of this nature will probably also have conquered not only old age but also disease—which may have been the condition of the human race prior to the Flood. *The Flood and the fall of man are inextricably connected.*

It could also be argued that these figures represent the ages not of individuals but of many generations. In this sense, future historians may be puzzled by the apparently illogically long life span credited to a king of England called George, not knowing that there were six of that name dating from the eighteenth to the twentieth centuries. But even if we do take this point of view, why do the ages gradually shorten with the passage of time after the Flood?

Perhaps these life spans are not mythological or symbolic, after all. They may not have been as long as the mythological material suggests, but one thing stands out: *the mythologies all point to a long life span of individuals before the Flood, with a gradual decrease of life span afterward,* until we reach the well-known three score and ten of historical times. In spite of all our advances both culturally and medically, this is the average life span of members of our civilized communities today. In underdeveloped countries the average life span is considerably shorter.

Much research is being done at present in an effort to halt the aging process and extend the human life span. We are also at the dawn of the age of space travel. It may well be that the future of space-traveling techniques and the conquest of age are not only complementary but dual prerequisites for prolonged interstellar travel and exploration. If we assume that our distant ancestors were members of a space-traveling community, might they not have solved these problems before their arrival on earth? And if they were both disease-free and long-lived, why did they cease to manifest these qualities?

Our hypothesis may suggest an answer:

We have assumed that earth was at this time nearer to the sun, with a consequent greater evaporation of surface water, resulting in a high-altitude water-vapor "screen" surrounding earth. This would have had the effect of a deeper, denser atmosphere. The surface pressure may have been slightly higher, with a consequent higher atmospheric pressure in high, mountainous regions. Another effect of this screen would have been to filter out a great deal of solar radiation, and certainly cosmic and hard X-radiation from extrasolar regions of the galaxy.

High-altitude layers of ice crystals have been detected in the upper atmosphere of Venus, which is nearer the sun than earth, so the assumption is not unreasonable.

Cosmic rays, it has been thought, may have a harmful effect on the human organism, which may be a factor both in the aging process and in certain diseases, particularly cancers. Some authorities have even suggested that a "deluge" of radiation reaching earth from a supernova explosion may have been responsible for the plagues of the past, some of which are unexplained. We do know that a great many deaths occurred at a particular time, which may have been caused by the bombardment of earth with high-intensity radiation from a specific source for a limited period.

If earth was shielded by a denser atmosphere from the harmful effects of radiation, a race of extrasolar colonists, who had conquered disease and extended their life spans, would not have been adversely affected by their new terrestrial environment. Indeed, if the conditions had not been satisfactory, either they would not have settled here in the first place or they would have taken certain steps to see that the conditions were altered to suit their needs.

This existence of an original human race enormously superior to ourselves would account for the widely diffused mythology of the "gods," the "godlike" men, the "mighty men of old," who lived for great periods of time. It would also account for legends of the Paradise World, the Golden Age, the Perfect World, which came to a sudden and calamitous end:

• The Garden of Eden was perhaps not so much a specific place as the *condition of the world before the catastrophe.*

• The Sumerians have the same legend: the land of Dilmun, pure, clean, and bright, where there is no sickness or old age.

Other mythologies describe the cities where the gods lived:

• In Norse mythology, there is Asgaard, located in Midgard, the "Middle of the Earth." From a high place called Hlidskjalf, the Norse god Odin could see all the world and the actions of men.

• The Aztecs have the legend of the bright city of Tula, or Tollan, of which the Aztec historian Tezozomoc said, "Where the bright sun lives, and where the god of light forever rules as long as that orb is in the sky." (The biblical New Jerusalem is described in almost identical terms.)

• Tula is also mentioned in the annals of the Maya, the Kichos, and Cukiliquels of Guatemala. The Maya book of *Chilam Balam* and the *Popol Vuh* have much to say about Tula: "Thence came we forth together, there was the common parent of our race."

When the Spaniards first arrived in the Americas they were told of bright cities, lit by stars, hung from the roofs, that never went out. The description reminds us of some form of electrical illumination, which was of course unknown in Europe at this time. No trace of cities with the remains of such means of illumination has been found. But among the ruined cities of the Maya, many large buildings have been found which are completely windowless but show no sign of blackening by torches, fires, or lamps.

Von Däniken has suggested that the "water conduits" at Tiahuanaco are so oddly situated and designed that they could not have been ducts for water, but were in fact the shielding for power sources.

From Iceland to South America legends talk of brightly lit cities —all of which were destroyed in some vast disaster. It is surprising how many curious legends we have, all of which represent things that are either in existence now or are not beyond our present scientific potential:

• The *philosopher's stone,* for example, is a concept of great antiquity. Transmutation, although not practicable chemically, would be feasible by the use of controlled nuclear reactions. If methods could be devised, it would be possible to change a metal—or any other substance—into another by rearranging their atomic structures. Passages in the Indian epic *Mahabharata* allude to such transmutations. Nuclear energy may have been known to the civilization of the remote past.

• The Arabian Nights story of Ali Baba features a *door that would open only when certain words were spoken.* It would not be difficult to install a device which would operate a lock in accordance with certain sonic vibrations, in this case, certain words. (We already have the photoelectric cell, by means of which the breaking of an invisible light beam switches in a circuit to operate a door mechanism.)

• What about the *flying carpet?* Could this be an echo of some sort of flying machine or platform? If man had always been primitive and had considered flying at all, he would surely have tried to copy the birds. But the concept of a flying *device* that does not resemble anything in nature is surely based on past experience.

• And what of the *magician's wand,* whereby many miraculous things were accomplished? Is this again the memory of a long-vanished device? The magician's wand is a common legend throughout the Middle East; it is also mentioned frequently in the Bible (the contest between the Egyptian magicians and Moses).

In the legends of South America, the culture god Wirakocha caused hills to be leveled, and buildings were erected "in a single night," with the aid of something like the magician's wand. On one occasion, he also used his wand to cause a fire to alarm some hostile natives.

Let us return to the philosopher's stone of the ancients, a legend that we have not been able to trace to its source. It is known that medieval alchemists attempted transmutation chemically. Theoretically, the only practical way to transmute one substance into another is to apply atomic energy; that is, to transform atomically.

The difference between elements lies in the composition of their atomic structures. For example, the atomic structure of gold is exactly the same as that of mercury, except that mercury has one more electron. If we remove one electron from each atom of mercury, we have an atom of gold. This has been demonstrated experimentally in the laboratory, but it has not yet been possible to envisage large-scale controlled reactions which would make such experiments commercially viable.

The exploitation of such a technique would mean that there would be no shortage of any needed materials. All the so-called valuable minerals—gold, silver, platinum, and gemstones—would be rendered worthless, as they could be produced in unlimited quantities. In fact, there would be no shortages of *any* material substances; even certain foods could be produced by this method. It is a step from transmutation to nuclear replication, by which an object could be copied by having its entire atomic and molecular structure reproduced.

Cultures of the very ancient past seem to have attached no great value to gold or gemstones. Gold was highly regarded because it was easily worked, and since it did not deteriorate when exposed to weather, it was used frequently to sheath buildings. The ancients attached no monetary value to these materials, as the Spaniards found when they invaded the Aztec and Inca territories.

Could it be that the peoples of these ancient cultures had no concept of financial systems or monetary values, because they themselves descended from a civilization in which such things could be produced at will and therefore had no intrinsic value? A world that could produce gold or anything else at will would have no particularly valuable minerals, and hence, no concept of money or personal wealth.

The "fountain of youth," the dream of extending youthful vigor indefinitely, may once have been a reality. We remember the long life spans mentioned in the Old Testament and other legends. These may be an echo of a time when humanity lived for vastly longer periods than we now do. There is actually no biological reason why human beings die at the comparatively early age of 60 to 70 years. Separated cell tissues have been kept alive under suitable conditions for an indefinite period.

The human brain has a capacity to store information at least a *million* times greater than that which is actually utilized. The human intelligence potential, in fact, seems to have been designed for a creature with a *much longer life span* than we possess today. It is doubtful that nature would equip any creature with an organ or potential *which has yet to be realized*. If intelligence were actually increasing, the brain would gradually increase in size and complexity over the generations, and would not possess, as it now does, large areas which are dormant or nearly so.

Although there is some evidence for a physical degeneration of the human species, it may appear that the greatest degeneration has taken place within human intellectual capabilities. This may account in part for the huge gaps of knowledge regarding many events in the remote past, and even of our true origins.

It would seem that if the human brain was meant to be utilized to a far greater extent than it is at present—even if it did not use its *full* potential—then the individual would necessarily have had a far longer life span than he does, to obtain the fullest possible benefit from such an increase of intellectual ability.

With longer life spans and vastly increased brain power—augmented by knowledge brought from a Superior Community elsewhere in space—the human race in the past would have been capable of scientific and inventive reasoning far beyond our present capacity. It does not necessarily follow, however, that such intellectual capacity would be complemented by perfect moral and spiritual values. Many of our faults may have existed within these advanced people, so that the seeds of their own destruction were already there. Because of their great intelligence, the inherent dangers were also greater.

If we assume that such people had created a civilization vastly superior to our own, it also follows that because of the difference in their mental capacity, this civilization may have followed a very different path from ours. It may have contained elements that are incomprehensible to us and excluded much of what we consider essential.

We have suggested that the cave-painting people from what we call Neolithic times may have had eidetic memory, or total recall. A race of people who possessed this faculty would have little need for our present aids to memory. All means of storing information artificially, from printed word to computer data banks, enable us to have at our disposal a great deal of information that our imperfect memories are incapable of storing "naturally." While a race with perfect recall would not need these aids, it is possible that they stored information in central places, or with certain classes of people, such as librarians. The information may have been stored in written, electronic, or other methods presently unknown to us.

By the same token, if they had managed to harness magnetic or gravitational fields on a large scale, they would have had little need for the cumbersome machines we use today. Both enormous weights and vast amounts of material could have been transported from place to place without the need for the vehicular transport devices we use today. These, of course, are speculations—but they are also possibilities. We may mention several more:

• Cutting tools may have used laser or ultrasonic power. Power may have been transmitted without cables and complex wiring.

• In view of certain information we now possess, it seems likely that they had limitless supplies of atomic energy.

We can only speculate on what form this civilization may have taken. Apart from the legends, we have only a little to go on. That little is encouraging, however, not so much for the things themselves as for their implications.

First, the maps:

• In 1929, an ancient map was discovered in Istanbul. It had been drawn by Piri Reis, grand admiral of the Turkish fleet, in 1513. The map, according to the admiral, had been drawn from information supplied by one of Columbus' sailors, as well as from maps dating back to the time of Alexander the Great.

Experts were amazed that the map showed the Brazilian and Argentine coasts, which by 1513 had not been charted by New World explorers. The map also showed lands to the south (the Antarctic continent) for which there were no corresponding areas in 1929.

During the International Geophysical Year, the Antarctic continent was charted by means of seismic soundings to establish the configuration of the land mass under the ice cap. When this was done, the new map was compared to Piri Reis' 1513 map. From the ancient maps Piri Reis had drawn part of the coast of Antarctica (the Queen

Maude area in particular). Certain discrepancies were noted between the two maps, and the seismic soundings were checked. It was found that the survey had been in error. *The ancient map was more accurate than the modern one.*

Even more startling, the Piri Reis map shows certain distortions of the American continent which occur only if earth is seen as a sphere from the air. Photographs taken by an orbiting satellite at a height of sixty miles showed the *same degree of distortion* as was recorded on the ancient maps.

The 1513 map could not possibly have been faked. When it was discovered in 1929, no one was aware of the shape of the Antarctic continent. A range of mountains shown on the map was not discovered until the International Geophysical Year.

The authenticity of the map has been recognized by the Royal Geographical Society and the American Geophysical Society.

• Ancient Nordic maps found among religious relics in the Icelandic cathedral show Greenland as three separate islands. The French polar expedition (1947–1949) led by Paul-Emile Victor undertook a seismic survey of Greenland, which showed that under the ice cap Greenland actually was composed of three separate islands. In shape and area, the ancient maps were extremely accurate.

Several conclusions can be drawn from these maps:

1. They must have been drawn before the present polar caps were formed.

2. Even without considering the possibility of high-altitude surveying and photographing in the case of the Piri Reis map of Antarctica, it is clear that very sophisticated ancient peoples must have mapped these now frozen lands. The high degree of precision shows that they must have used accurate surveying equipment. Even the highly civilized Greeks and the Romans drew maps of the known world by rule of thumb, guesswork, and the knowledge drawn from coastal voyages. Although they show approximate dispositions of major land masses, they are inaccurate in detail and the shape of various lands and islands. Maps made as late as the fifteenth and sixteenth centuries were just as inaccurate.

The origin of astrology, one of the most ancient sciences, is another puzzle. How did men first arrive at the idea that certain groups of stars could possibly influence human destinies?

Michael Auphan, in his book *L'Astrologie par la Science* (1956), has proposed that the sun emits as yet unknown rays, which he calls odique waves. These waves, which follow the laws of electromagnetic fields, penetrate earth. Earth acts like an oscillating sphere bathing in a magnetic field.

In developing his idea, Auphan has theorized mathematically that there would be twelve sectors of magnetic influence on the ecliptic which correspond to the twelve signs of the zodiac. Man, he says, is affected by the magnetic interaction between the earth and the sun. The originators of astrology, he suggests, were aware of this influence, which has since been distorted into the modern art of astrology. *These ancient astrologers would have to have had a knowledge of physics greater than that possessed by scientists today.*

12
THE
ATLANTIS HYPOTHESIS

If humanity has occupied the earth for a period of 30,000 years or so, and we have formulated theories that civilization started only some 6,000 years ago, what had happened to the intervening ages?

Why, after over 20,000 years of apparent barbarism, was there a sudden rash of city-building with all it implies—mathematics, plant and animal husbandry, art and architecture, government and law?

If human history unfolded in this way, then something sudden and dramatic must have happened 6,000 years ago. Some impetus must have jolted the human race out of its animal rut, and if this impetus did not exist within the human race, then it must have had an external source.

Von Däniken, in his book *Chariots of the Gods?*, postulates that visitors from space taught our ancestors the rudiments of civilization and then departed, leaving a memory of gods. This theory would certainly explain some of the peculiarities associated both with ancient civilizations and with the development of religion. Its major difficulty, however, is that it does not take into account the mythology of a Golden Age, a great civilization which existed prior to the Flood.

The people described as the culture bearers, who brought civilization to the survivors of the Flood, appear themselves to have been flood survivors. They owed nothing to an extraterrestrial origin or extraterrestrial influence.

But if a superior civilization existed prior to the Flood, what was the nature of the catastrophe that destroyed it? It seems that the catastrophe and the Ice Age concept are linked. The Ice Age theory, we have tried to show, cannot account either for the legends or for the physical evidence of widespread destruction.

In considering these rather knotty questions, archeologists generally agree on two things:

1. The suddenness with which urban civilizations appeared in many parts of the world.

2. The intimate connection between the culture bearers and the rise of civilization.

They have not, however, determined either the origin of these culture bearers or the source of the knowledge with which the culture bearers re-created a civilization.

Did these civilizers learn their arts from a pre-Flood civilization?

Plato's description of the legendary kingdom of Atlantis has assumed momentous importance in discussing the origin of civilization.

All the ingredients are there: a high civilization, centuries before the rise of the Greeks or the Egyptians, destroyed by a great catastrophe. The most important element of the Atlantis legend, however, is its location. It is assumed to have been situated in the Atlantic Ocean between Europe and America.

Experts have been trying for many years to discover why so many similarities exist between Old and New World civilizations—Sumerian, Egyptian, Maya, Pre-Toltec, Roman, and Inca. Plato's discourse on the legend of Atlantis has been cited as the "missing link" between these diverse cultures.

Interpretations of the Atlantis legend emphasize its location in the Atlantic Ocean beyond Gibraltar near the Azores. From here it is a short distance to the entire Mediterranean basin area and the Middle East. The distance across the Atlantic Ocean is not too great; the Gulf of Mexico and the principal areas of the ancient cultures of Middle America lie on a course slightly southwest. From here, migrations across the Isthmus of Panama would bring travelers to the Pacific coast of South America, the area of the Inca, Chimu, Mochica, and Chavin cultures.

The adherents of the Atlantis hypothesis make much of the fact that in the Old World there are step pyramids, from the Egyptian and the Sumerian cultures which have apparent parallels in the complex of Maya remains, including El Castillo at Chichén Itzá, and the pyramids at Tikal and Palenque. The Pyramid of the Sun at Teotihuacán in Mexico actually exceeds in area the Great Pyramid of Cheops in Egypt. Both Old and New World civilizations share the legends of the Flood and the culture bearers, and it has also been claimed that there is a great deal of similarity between the Inca language Quechua and ancient Sumerian.

These similarities and many more, including art forms, have been cited as evidence that Plato's Atlantis was in fact the original motherland of the cultures of antiquity.

There are, however, difficulties with this thesis. Let us examine the Atlantis legend more closely. In doing so, we may both clarify these discrepancies and discover how the Atlantis mystery fits in with other events and legends also misconstrued as parts of the Flood theory.

Around 360 B.C., Plato wrote two works, *Timaeus* and *Critias*, which include the story of Atlantis. The story was first told to Plato's ancestor, Solon, the Greek lawgiver (638–559 B.C.), by the Egyptian priests of Sais, a city which at that time had close ties with the Athenians.

This was the story the priests told Solon, as recorded in *Timaeus:* "Our histories tell of a mighty power which unprovoked made an expedition against the whole of Europe and Asia, and to which your city put an end. This power came forth out of the Atlantic Ocean, for in those days the Atlantic was navigable. There was an island situated in front of the Straits which are by you called the pillars of Herakles. The island was larger than Libya and Asia put together, and was the way to other islands, and from these you might pass to the whole of the opposite continent which surrounded the true ocean.

"Now in this island of Atlantis there was a great and wonderful empire which had rule over the whole island and several others, and over parts of the continent. The men of Atlantis had subjected the parts of Libya within the Pillars of Herakles as far as Egypt, and of Europe as far as Tyrrhenia. This vast power, gathered into one, endeavored at a blow to subdue our country and yours and the whole of the region within the Straits.

"And then, Solon, your country shone forth, in the excellence of her virtue and strength. She was pre-eminent in courage and military skill, and was the leader of the Hellenes. And when the rest fell off from her, being compelled to stand alone, after having undergone the very extremity of danger, she defeated and triumphed over the invaders, and preserved from slavery those who were not yet subjugated, and generously liberated all the rest of us who dwell within the Pillars.

"But afterward there occurred violent earthquakes and floods; and in a single day and night of misfortune all your warlike men in a body sank into the earth, and the island of Atlantis in like manner disappeared in the depths of the sea. For which reason the sea in those parts is impassable and impenetrable, because there is a shoal of mud in the way; and this was caused by the subsidence of the island."

And from the *Critias:*

"And Poseidon, receiving for his lot the island of Atlantis, begat children by a mortal woman, and settled them in part of the island. Looking toward the sea, but in the center of the whole island, there was a plain which is said to have been the fairest of all plains and very fertile. Near the plain again, and also in the center of the island at a distance of about fifty stadia, there was a mountain not very high on any side. In this mountain there dwelt one of the earth-born primeval men of that country, whose name was Evanor, and he had a wife called Leucippe, and they had only one daughter who was called Cleito.

"The maiden had already reached womanhood when her father and mother died. Poseidon fell in love with her and had intercourse with her, and breaking the ground, enclosed the hill in which she dwelt all around, making alternate zones of sea and land larger and smaller, encircling one another. There were two of land and three of water, which he turned as with a lathe, each having its circumference equidistant every way from the center, so that no man could get to the island, for ships and voyages were not as yet.

"He himself, being a god, found no difficulty in making special arrangements for the center island, bringing two springs of water from beneath the earth, one of warm water and the other of cold, and making every variety of food to spring up abundantly from the soil. He also begat and brought up five pairs of twin male children. Dividing the island of Atlantis into ten portions, he gave to the firstborn of the eldest pair his mother's dwelling and the surrounding allotment, which was the largest and best, and made him king over the rest. The others he made princes, and gave them rule over many men, and a large territory. And he named them all; the eldest, who was the first king, he named Atlas, and after him the whole island and the ocean were called Atlantic."

Plato described the island of Atlantis in great detail:

"First of all they bridged over the zones of sea which surround the ancient metropolis, making a road to and from the royal palace. And at the very beginning they built the palace in the habitation of the god and of their ancestors, which they continued to ornament in successive generations, every king surpassing the one who went before him to the utmost of his power, until they made the building a marvel to behold for size and for beauty.

"And beginning from the sea they bored a canal for three hundred feet in width and one hundred feet in depth and fifty stadia in length, which they carried through to the outermost zone, making a passage from the sea up to this, which became a harbor, and leaving an opening sufficient to enable the largest vessels to find ingress. Moreover, they divided at the bridges the zones of land which parted the zones of sea, leaving room for a single trireme to pass out of one zone into another, and they covered over the channels so as to leave a way underneath for the ships, for the banks were raised considerably above the water.

"The island in which the palace was situated had a diameter of five stadia. All this including the zones and the bridge, which was the sixth part of a stadium in width, they surrounded by a stone wall on every side, placing towers and gates on the bridges where the sea passed in. The stone which was used in this work they quarried from underneath the center island, and from underneath the zones, on the outer as well as the inner side. One kind was white, another black, and a third red, and, as they quarried, they at the same time hollowed out double docks, having roofs formed out of the native rock. Some of their buildings were simple, but in others they put together different stones, varying the color to please the eye, and to be a natural source of delight. The entire circuit of the wall, which went round the outermost zone, they covered with a coating of brass, and the circuit of the next wall they coated with tin, and the third, which encompassed the citadel, flashed with the red light of orichalcum.

"They had fountains, one of cold water and another of hot water, in gracious plenty flowing; and they were wonderfully adapted for use by reason of the pleasantness and excellence of their waters. They constructed buildings about them and planted suitable trees; also they made cisterns, some open to the heavens, others roofed over, to be used in winter as warm baths. There were the king's baths, and the baths of private persons, which were kept apart; and there were separate baths for women, and for horses and cattle, and to each of them they gave as much adornment as was suitable.

"Of the water which ran off they carried some to the grove of Poseidon, where were growing all manner of trees of wonderful height and beauty, owing to the excellence of the soil, while the remainder was conveyed by aqueducts along the bridges to the outer circles. There were many temples built and dedicated to many gods; also gardens and palaces of exercise, some for men, and others for horses in both of the two islands formed by the zones; and in the center of the larger of the two there was set apart a racecourse of a stadium in width and in length allowed to extend all around the island, for horses to race in.

"Leaving the palace and passing out across the three harbors, you came to a wall which began at the sea and went all around. This was everywhere distant fifty stadia from the largest zone or harbor and enclosed the whole, the ends meeting at the mouth of the channel which led to the sea. The entire sea was densely crowded with habitations; and the canal and the largest of the harbors were full of vessels and merchants, coming from all parts, who, from their numbers, kept up a multitudinous sound of human voices, and din and clatter of all sorts night and day.

"The whole country was said to be very lofty and precipitous on the side of the sea, but the country immediately about and surrounding the city was a level plain, itself surrounded by mountains which descended toward the sea. It was smooth and even, and of an oblong shape, extending in one direction three thousand stadia, but across the center island it was two thousand stadia. This part of the island looked toward the south, and was sheltered from the north. The surrounding mountains were celebrated for their number and size and beauty, far beyond any which still exist, having in them also many wealthy villages of country folk, and rivers, and lakes, and meadows supplying food enough for every animal, wild or tame, and much wood of various sorts.

"The plain was for the most part rectangular and oblong, and where falling out of the straight line followed the circular ditch. The depth, and width, and length of the ditch, were incredible, and gave the impression that a work of such extent, in addition to so many others, could never have been artificial. Nevertheless I must say what I was told. It was excavated to the depth of a hundred feet, and its breadth was a stadium everywhere. It was carried around the whole of the plain, and was ten thousand stadia in length. It received the streams which came down from the mountains, and winding around the plain and meeting at the city, was there let off into the sea.

"Further inland, likewise, straight canals of a hundred feet in width were cut from it through the plain, and again let off into the ditch leading to the sea. These canals were at intervals of a hundred stadia, and by them they brought down the wood from the mountains to the city, and conveyed the fruits of the earth in ships, cutting transverse passages from one canal into another, and to the city. Twice in the year they gathered the fruits of the earth—in winter having the benefit of the rains from heaven, and in summer the water which the land supplied by introducing streams from the canals."

Plato mentions the population, the military forces, which consisted of 10,000 chariots and 1,200 ships of war, the laws, and the administration. The people of Atlantis remained great, he said, as long as the divine element lasted in them.

"When the divine portion began to fade away, and became diluted too often and too much with the mortal admixture, and the human nature got the upper hand, they then, being unable to bear their fortune, behaved unseemly, and to him who had an eye to see, grew visibly debased, for they were losing the fairest of their precious gifts; but to those who had no eye to see the true happiness, they appeared glorious and blessed at the very time when they were full of avarice and unrighteous power.

"Zeus, perceiving that an honorable race was in a woeful plight, and wanting to inflict punishment on them, that they might be chastened and improved, called all the gods together to address them."

The story of Atlantis was never completed by Plato, and in fact breaks off in mid-sentence.

Largely through this account, Atlantis has become the subject of a great many hypotheses. It has been located in Spain, Britain, Iceland, Scandinavia, Brazil, Yucatán, and countless other locations. Apart from the Mediterranean, the location most often theorized was the Azores, which fit most closely with Plato's description. The Pillars of Hercules to which he referred may be a reference to Gibraltar. Furthermore, the Azores are of volcanic origin, consist of the same geological formations mentioned by Plato, namely red, white, and black rock, and contain hot and cold springs.

Although small volcanic islands may have risen and sunk within recent times, however, it has never been proved that a large island mass has existed in recent geological times in the vicinity of the Azores. Cores from the bottom of the Atlantic near the Azores show evidence of vulcanism of a pattern characterized only by land volcanoes. This may more reasonably suggest small volcanic islands than a large land mass that sank.

The main problems in solving the Atlantis riddle are the time factor and the size of the island described by Plato.

First, the time factor: Plato dates the disaster that destroyed Atlantis 9,000 years before Solon's time, or 12,000 B.C. A high culture *could* have existed at such an early date, but Plato's dates do not conform to other evidence.

The Atlantis Plato described is very similar to a Mediterranean type of culture not noticeably different from that readily understood by Solon and Plato. There is also a mention of a war between Atlantis and the Greeks, and the descriptions of chariots and galleys. The 9,000 years is perhaps not an exaggeration but an error in dating.

Second, if we are considering an island-based empire in the Mediterranean, the size of the island or islands constitutes a further problem.

Many authorities have reasoned that if Plato's Atlantis was based on historical fact, it referred to the island-based Minoan sea-trading empire of Crete. The difficulty in this thesis was in reconciling the statistics given by Plato with the geography of Crete and its associated islands.

The size of the plain, as given by Plato, was three thousand by two thousand stadia, or roughly 340 by 230 miles. This plain is much larger than the plain of Messara in Crete, or of any plain on mainland Greece.

The ditch was stated to have been 10,000 stadia or 1,100 miles long, and was divided into 60,000 lots of land, each one square mile in area. The leader of each lot was required to furnish for the war between Greece and Atlantis one sixth of a war chariot, two horses and riders,

The Greek stadium is 600 Greek feet, and it has recently been discovered that the length of the foot in Minoan Crete was almost exactly the same as the present English foot.

one light chariot, a foot soldier with shield, a charioteer, two heavily armed men, two archers, two slingers, three stone shooters, three javelin men, and four sailors to man the ships, of which there were 1,200. According to this formula, the military forces of Atlantis would have comprised 1.2 million men, an army far larger than any Bronze Age civilization could have possibly mustered.

But what if Solon had erroneously translated the symbol for 100 as 1,000? The two symbols in Cretan script are almost identical. (Dr. Angelos Galanopolous, a Greek seismologist, first suggested this error in translation as a means to resolve the problem of dates.) Instead of 9,000 years, we would have 900 years before Solon. The disaster would have occurred about 1500 B.C. If the size of the plain were reduced by a factor of ten, to 34 by 23 miles, it would approximate closely the size of the plain of Messara on Crete.

It has been reckoned that the Royal State of Atlantis, and the Citadel, or Capital, refer actually to two islands; the larger would be Crete, and the Island of the City would be Thera. Again, reduced by a factor of ten, the dimensions given by Plato fit Thera almost exactly.

Reduced by ten, the 60,000 lots become 6,000; 1,200 ships become 120 ships, and the size of the army is reduced to 120,000 men, which could conform with the kind of military power exercised in the Mediterranean in the second millennium B.C.

These figures would seem to make sense in connection with the Minoan sea empire of Crete, with its capital city on Thera. Further, it has recently been found that around 1500 B.C., a colossal volcanic eruption occurred on the island of Thera, which completely destroyed the center of the island. The ensuing tidal waves, earthquakes, and deposits of volcanic ash wrought havoc throughout the entire Mediterranean basin, Egypt, the Palestine coast, Turkey, and mainland Greece, and virtually destroyed the civilization of Crete.

The volcanic eruption on Thera was, according to geologists' best estimates, at least four times more powerful than that of Krakatoa in 1883, which itself was equivalent to the simultaneous detonation of

430 H-bombs. The explosion on Thera, therefore, would have been equivalent to the explosion of 1,500 to 2,000 H-bombs, and the devastation, with its consequent effects on the peoples of the entire Mediterranean area, must have been on a scale which has had no parallel since.

As the eruptions on Thera in 1500 B.C., and Krakatoa in 1883, were of the same type, it has been possible to reconstruct the effects of Thera's eruption, using Krakatoa as a model. When the island of Krakatoa was destroyed, it disappeared in twelve hours. (Plato says that Atlantis sank in a day and a night, a period of twenty-four hours.) Deposits of reddish ash fell over thousands of square miles, and three great seismic sea waves caused havoc in Java and Sumatra, drowning thousands of people. The amount of ash suspended in the atmosphere caused days of darkness over thousands of square miles, and spectacular sunsets all over the world for years afterward. The sound of the explosion could be heard in Australia, over a thousand miles away.

We could expect similar effects from the Thera eruption of 1500 B.C., but on a much larger scale. Many legends, including much Old Testament biblical material, tie in closely with such volcanic phenomena.

In Greek mythology, Zeus, angered by the behavior of the sons of Lycaon, is determined to destroy humankind. Prometheus warns his son, Deucalion, of the impending disaster, who accordingly builds an ark to ride out the flood. A terrible south wind springs up, the rain falls, and the rivers roar down to the sea which, rising with astonishing speed, wash away every city of the coast and plain. The entire world, except for a few mountain peaks, is flooded and all mortal creatures seem to have been lost, except Deucalion and his wife Pyrrha. Deucalion's ark floats about for nine days, until at last the waters subside. The legend states that there were three floods.

In 1958, an earthquake in Alaska raised seismic waves more than 1,200 feet high, a phenomenon called by the Japanese name tsunami. The Thera eruption could have caused greater waves than this, and it may well have appeared to the Greeks that the whole world had been drowned by walls of water.

The events described in Exodus have long been thought to have been connected with a volcanic eruption. The problem is, where could such a volcano have been situated? The Middle East is singularly devoid of volcanic mountains of recent geological origin. However, the Thera disaster was on a scale that would have been felt throughout the Middle Eastern region.

The Hebrews' exodus from Egypt took place during what were called the plagues, at the end of the Middle Kingdom, at almost the same time (ca. 1500 B.C.) the Hyksos were invading Egypt. No doubt the chaos created by a volcanic eruption like that of Thera would have provided ideal conditions for both the Hebrews' exodus and the Hyksos' invasion.

The twelve plagues of Egypt follow this pattern:

1. Serpents
2. Waters turned to blood
3. Frogs
4. Lice
5. Flies
6. Murrain of animals
7. Boils
8. Pestilence
9. Hail, thunder, and fire
10. Locusts
11. Darkness
12. Death of the firstborn

The first phase of the Thera eruption was the deposit of rose-colored pumice. In the case of Krakatoa, this phenomenon was observed at 7 A.M. in Djakarta, on Java, 100 miles from Krakatoa; the sky began to darken and become yellow. By 11 A.M. there was a heavy dust rain followed by complete darkness, which lasted until after 3 P.M. more than 150 miles from Krakatoa. The "waters turned to blood," therefore, were caused by the deposits of red pumice from the eruption.

This effect on the waters would have caused the frogs to leave the water and die on land, which would subsequently bring about the plagues of lice and flies to feed on their decomposing bodies. The ash would also cover all the plant life and kill it, so that the animals would be unable to feed, which would be the "murrain of animals." The boils and pestilence refer to bubonic plague and other sicknesses, probably cholera and typhoid, brought about by the lack of food and polluted drinking water.

The hail was the rain of stones and ash, mixed with hail: "Throughout Egypt the hail struck everything in the fields, both men and beast; it beat down every growing thing and shattered every tree." (Exodus 9:24)

During the severe eruption of the type characterized by Thera and Krakatoa, the eruption is accompanied by thunder and lightning; in the case of Krakatoa, the rain of mud, lava, and hail was phosphorescent, and in the intense darkness gave an impression of being on fire. Exodus 9:23 reports, ". . . and the Lord sent thunder and hail, with fire flashing down to the ground."

The plague of locusts was a direct result of the eruption. These creatures would be escaping from areas affected before Egypt, and were in the process of migration.

The "death of the firstborn" is the only plague that does not fit in with an explanation of the plagues as natural phenomena. Either this final plague was added later as Hebrew propaganda, or a misunderstanding has arisen. Some authorities have stated that "firstborn" is a mistranslation of a Hebrew word *bkhor* (firstborn), whereas it should have been *behor* (chosen).

According to this "revised" translation, not one house of the "chosen" (Egyptians) escaped without a death. This may have been due in part to the fact that the Israelites were a separate community in Egypt at this time; in addition, because of their relatively smaller numbers, they were not as affected by infectious diseases. However, there may be a certain degree of propaganda involved in the Exodus story; Velikovsky has stated that there are extant scrolls (not included in the final version of Old Testament Exodus) that report that *three quarters of the Hebrews died of suffocation* during the rain of ash and stones.

Regarding the actual exodus of the Israelites from Egypt, Exodus 14 has this to say:

"And Moses stretched out his hand over the sea; and the Lord caused the sea to go back by a strong east wind all that night and made the sea dry land, and the waters were divided.

"And the children of Israel went into the midst of the sea upon the dry ground; and the waters were a wall unto them on their right hand, and on their left.

"And the Egyptians pursued, and went in after them in the midst of the sea, even all Pharaoh's horses, his chariots and his horsemen.

"And it came to pass in the morning watch that the Lord looked into the host of the Egyptians through the pillar of fire and of the cloud that troubled the hosts of the Egyptians.

"And took off their chariot wheels, that they drave them heavily; so that the Egyptians said, Let us flee from the face of Israel; for the Lord fighteth for them against the Egyptians.

"And the Lord said unto Moses, Stretch out thy hand over the sea, that the waters may come again upon the Egyptians, upon their chariots and upon their horsemen.

"And Moses stretched forth his hand over the sea, and the sea returned to his strength when the morning appeared; and the Egyptians fled against it; and the Lord overthrew the Egyptians in the midst of the sea.

"And the waters returned, and covered the chariots, and the horsemen, and all the host of Pharaoh that came into the sea after them, and there remained not so much as one of them.

"But the children of Israel walked upon dry land in the midst of the sea, and the waters were a wall unto them on their right hand, and on their left.

"Thus the Lord saved Israel that day out of the hand of the Egyptians; and Israel saw the Egyptians dead upon the sea shore."

Scholars agree for the most part that the Israelites made their way out of Egypt not by the *Red* Sea but by the *Reed* Sea, or Yam Suf (it is thought that "Suf" means "reed"). A script found at El-Arish says that the reed sea was a lagoon or lake between Romani and El-Arish, east of the Nile Delta, and ran parallel with the Mediterranean to what today is called Kash Bouroun. This sea or lagoon was known in ancient times as the Serbonis Sea. Herodotus says that the Serbonis Sea, according to legend, was where the monster Typhon lived (this itself is a connection with the Thera eruption) and that it was dangerous to attempt a crossing of the sea. The length was given as 200 stadia, and the width 50 stadia, 20 by 5 miles.

It has been suggested that the number of Israelites was grossly exaggerated, and was probably nearer 600 than 600,000. The population of ancient Egypt at this time was probably only several million at most, and the influx of a minority group—which is what the Israelites were—numbering in the hundreds of thousands is most unlikely. This would have increased the population of Egypt by a third, a state of affairs the Egyptian authorities would hardly have tolerated. The resources of the country would hardly have supported such an increase in population size.

The seismic sea waves resulting from Thera's eruption could have drawn the water from the shallow lagoon for a period of time sufficient for at least most of the Israelites to have made their escape when the opportunity came. The returning wave trapped the Egyptian army. The biblical account says that the chariot wheels stuck, implying soft mud or quicksand; this was also mentioned in Herodotus' description of the Serbonis Sea.

The pillar of fire by night, and a pillar of cloud by day, contained in the biblical account, obviously refers to the plume from erupting Thera. The Israelites, crossing the Serbonis Sea, would have been facing the Mediterranean and would have seen such "pillars."

Much Old Testament writing is dominated by catastrophic events, most of which appear to be dated around 1500 B.C. These events, mentioned by many of the prophets of Israel, seem to have been inspired in large measure by the vast eruption of Thera and its consequences upon life in the Mediterranean area. But the Old Testament is not the only source reporting the effects of a vast catastrophe of volcanic origin in the eastern Mediterranean:

• The story of the Argonauts of Jason contains passages which appear to refer to some great eruption in the Mediterranean. Apollonius says: "The next night caught them well out in the wide Cretan Sea, and they were frightened, for they had run into that sort of night that people call the pall of doom. Black chaos had descended on them from the sky, or had this darkness risen from the nethermost depths? They could not tell whether they were drifting through Hades or still on the waters."

• A papyrus written by Ipuwer, who must have been an eyewitness, contains a description of the effects of an earthquake in the wake of the eruption: "The towns are destroyed. Upper Egypt has become a waste. All is ruin. . . . The Residence is overturned in a minute."

• A stele at El-Arish bears this hieroglyphic inscription: "The land was in great affliction. Evil fell on this earth. There was a great upheaval in the Residence. . . . Nobody could leave the palace during nine days, and during these nine days of upheaval there was such a tempest that neither men nor gods could see the faces of those beside them."

• Exodus 10:22: "And there was a thick darkness in all the land of Egypt three days. They saw not one another, neither rose from his place for three days."

All these descriptions obviously refer to the effects of the eruption—earthquakes and the pall of darkness created by the cloud of ash and dust from Thera.

Greek descriptions of the disaster have it emanating from the south, while the Israelites and the Egyptians note that the disaster originated in the north. This accords perfectly with an eastern Mediterranean origin.

Many of the events in the Old Testament refer without doubt to the great volcanic eruption of Thera. The book of Jeremiah, written in the sixth century B.C., for example, reports (47:2–4): "Behold, waters rise up out of the north, and shall be an overflowing flood, and shall overflow the land. Because of the day that cometh to spoil all the Philistines . . . the remnant of the country of Caphtor."

The reference to the "remnant of the land of Caphtor" refers, it is thought, to the island of Crete; the Philistines were descendants of the Minoan civilization who fled from their homes in the Aegean after the collapse of the Cretan empire following the catastrophe.

Caphtor was known as Keftui to the Egyptians. The cessation of all contact and trade with these people following the eruption may well have led the Egyptians to believe that the whole land had mysteriously sunk beneath the sea with all the inhabitants.

All the evidence, from Plato to more recent archeological findings, from various mythologies and the Old Testament, suggests that *Atlantis was actually the Cretan sea empire, catastrophically destroyed by the eruption of the volcanic island of Thera circa 1500* B.C.

We have gone into the problem of Atlantis in some detail for two reasons:

1. The many theories developed to prove that all civilization in both the Old and the New Worlds originated in Atlantis are in error.

2. They show how easily errors of large proportions can be made when theorizing about the effects of the cataclysmic disruption of human affairs, by telescoping separate events into one event or even one series of events. Thus, the Flood myths and the destruction of the Cretan empire (Atlantis) have been assumed to be one and the same thing. The Flood of Deucalion and the Flood of Noah are also thought to have been different aspects of the same disaster.

From examination of the biblical record alone, we now know that the Flood of Noah was a different catastrophe from the destruction of Atlantis—which forms the Exodus narrative. The Bible is quite specific on this point. The Flood, according to biblical chronology, happened several thousand years *before* the Exodus of the Israelites from Egypt.

What emerges most clearly from the foregoing is that even those mythologies and legends that seem most fantastic are *firmly based in reality.* (Plato's Atlantis has been shown to be true, even in its details of dimensions.) Perhaps if we relied more on interpreting the evidence contained in our mythologies than on some of the inconsistent theories of our experts, we might get much nearer to the truth.

The beginning of Plato's discourse reports that the kingdom of Atlantis arose suddenly and miraculously under the aegis of Poseidon; the archeological evidence shows that civilization blossomed suddenly on Crete, as in many other places, before 2000 B.C. If we could decipher the Cretan linear scripts, we should find, as in the case of the Sumerian clay tablets, a Flood epic similar to that of Gilgamesh, with the origin of Cretan civilization in a reign of gods or demigods.

We also know that the Flood legend is worldwide. It is not likely that all of these variations on the Flood theme stem from a single eruption, however great, in the eastern Mediterranean. Although trans-Atlantic contacts may have existed at this time, a story brought by travelers to the Americas and the Pacific would not have left such a great impression on the people in these regions. It seems more probable that the Flood that inspired New World legends like those of the Old World was an event New World inhabitants actually experienced, too.

If the Cretan empire *was* Atlantis, and, accordingly, the cradle of *all* civilization, in both the Old and New Worlds, then the American cultures of antiquity were younger than that of Crete. Evidence seems to show, however, that the ancient civilizations of the Americas are, for the most part, contemporary with those of the Old World. Some Maya centers, for example, go back to at least 2000 B.C., and there are traces of cities which may be much older than these. Some of the cultures of the coastal regions of South America may be as old as Sumer, Crete, and Egypt. The so-called Mound Builder cultures of the American Southwest also appear to be contemporary with the Megalithic cultures of Europe, which date to a period prior to the second millennium B.C.

Atlantis, then, was not the mother of civilization, but was itself part of the sudden emergence of civilization all over the world. A prior catastrophe destroyed a high culture and left the survivors to rebuild civilization.

Egypt, Sumer, Greece and Crete, Toltec, Maya and Inca were the cultures built on the wreckage of this earlier civilization. These pre-Flood people were the "gods," and the Flood survivors were the revered "culture bearers" of later ages.

13

WAS THE FLOOD A NATURAL DISASTER?

The Flood legend appears in the mythology of virtually every people in the world. According to these mythologies, this catastrophe destroyed most of the human race, completely exterminated many species of animals, vastly altered the climate of the planet, and changed both its axial inclination and its orbital distance from the sun.

From evidence already cited—including areas of human knowledge for which there appears to be no traceable, historical source, and the existence of accurate maps of regions of the earth's surface now buried under the ice, we have assumed an advanced state of civilization prior to the Flood.

What was called the Wurm glaciation has been estimated to have ended some ten thousand years ago. In Antarctica, however, glaciation appears to have commenced some *six* thousand years ago. We have theorized, from the evidence, that there was no Ice Age, or Ice Ages; that, on the contrary, *the present is the period of glaciation* due to a deterioration in the earth's climate.

133

From the evidence at hand, we can approximate a time scale for the date of the catastrophe. The upper limit can be set at 8000 B.C. and the lower limit at around 4000 B.C. It would seem that we could narrow it down to around 6000 B.C., for the first traces of settled urban communities date from that time: the oldest so far excavated are the town sites of Catal Huyuk and Jericho on the Anatolian plain in Turkey.

My original theory regarding the nature of the catastrophe, based largely on legendary accounts, involved a fire prior to the Flood. At that point, still firmly convinced that there were Ice Ages (and, for that matter, that the evolution of man was a reasonable theory), I assumed that a release of great heat would have melted the ice and led to the Flood. As it became increasingly evident that there was no Ice Age, this idea had to be discarded. However, other factors will lead to the same consequences.

If earth, as we have hypothesized, had at one time been nearer the sun, with a great deal of water vapor suspended in the upper atmosphere, a cooling of this vapor would still lead to a Flood. A "fire" like that described in worldwide legends would have initiated the cooling of the earth and hence the precipitation.

Consider, for example, a cometary mass of the type described by Velikovsky. The effect of the comet—or "fire"—upon the earth, by altering its orbit, would have led to the cooling of the atmosphere, with consequent flooding. Our surveys both of cometary constitution and of the behavior of large planetary bodies, however, seem to rule out the possibility of the collision of a comet with earth. We must seek another solution.

Let us refer to legends once again.

• Genesis 7:11: ". . . all the springs of the great abyss broke through, the windows of the sky were opened, and rain fell on the earth for forty days and nights." Not only did rain fall from the sky, but waters rose up out of a "great abyss" or rift in the earth, and waters came up onto the earth.

• The ancient Sanskrit text called the *Mahabharata*, a collection of books written in their present form about three thousand years ago (it is possible they were copies of still earlier books no longer in existence), contains this passage: "Then men tampered with the 'Divine Fire' so that the earth split apart and sixty million people in great cities were drowned in one terrible night."

• Attempts to decipher the Mayan *Troano Manuscript* led to this translation: "The country of the hills of earth, the Land of Mu, were sacrificed. Twice upheaved, it disappeared during the night, having been constantly shaken. The land rose and sank several times in various places, and at last the surface gave way and the ten regions were torn asunder and scattered, the millions of inhabitants sank also."

These three legends, one from the Old Testament of the Bible, originating with the Hebrews of the Middle East, another from the *Mahabharata,* a document from India, and the third from Central America, are extraordinarily similar in their contents.

All three describe the opening up of the earth; one (Genesis) speaks of most of humanity perishing without specifying numbers, the other two mention the death of millions by drowning.

These legends date from the very beginnings of civilization as we know it. They were already traditions *before* these civilizations arose, and hence must predate our earliest records of these civilizations.

We have been led to believe that the total world population in Neolithic times, before the rise of urban cultures, was approximately 5 million, and progressively smaller the further we go into the past. At the same time, these legends speak of populations of great proportions, great cities with millions of inhabitants. (The *Mahabharata* says "sixty million people perished in one terrible night.")

We are given the impression from these ancient documents that the world was at one time densely populated and that most of the people were destroyed. At the beginning of Genesis, for example, it is said that men filled the earth. *After* the catastrophe, it would not be unreasonable to say that out of a world population of several hundred million people, only 5 million survived.

We cannot be sure that populations of hundreds of millions did not exist. Decrying ancient texts is no answer; as we have seen from Plato's Atlantis, legend contains a great deal of truth. What has been lost in the intervening centuries is the concrete evidence.

The worldwide flood legends—and this is true of the Hebrews, the Sumerians, Indian tribes of the Americas, the Polynesians, and the peoples of Southeast Asia—are remarkably unanimous in their main themes. A man and a woman are warned by the god that the world is to be destroyed, and they must build a boat and escape the flood in this way, together with such animals and foods as they are able to take with them.

We note that in Genesis, for example, the Lord warned Noah to prepare the Ark, and take his wife and family and male and female of each animal, and that he had seven days in which to do this before the Flood came.

Taking these legends as a whole, we are left with the picture of many human groups surviving in boats, perhaps thousands of boats, and that they had some warning of the catastrophe.

The Old Testament version of the Flood contains this reference to the building of the Ark: "The length of the Ark shall be three hundred cubits, its breadth fifty cubits, and its height thirty cubits."

The Hebrew cubit is usually estimated to have been about 1.6 feet; therefore, the length of the Ark was 480 feet, its breadth 80 feet, and the height 48 feet. Even by modern standards this would be large, particularly for a wooden ship. With these proportions, it was certainly much larger than those built in Europe at the time of Columbus.

If there were only a few days' warning of what was to happen, a considerable labor force, working at great speed, would have been required to build the vessel, supply it, and board and load it within the allotted time. Seven days seems hardly long enough; perhaps the warning came more than seven days in advance. We can be reasonably sure, however, that there was a period of warning, and that it was short.

At the end of the so-called Pleistocene glaciation, many millions of animals were killed violently and suddenly. Some species—the mammoth, the woolly rhinoceros, the giant elk, the American elephant, the giant sloth, and the cave bear—were completely eliminated, and have become extinct. Remarkably, all of these species were especially large and possibly dangerous animals. Domesticated and smaller species of animals have survived to this day.

Let us pursue this idea further. We have noted that there must have been many ships, in many parts of the world, built to escape the Flood. These ships carried cargoes of animals as well as people. This would seem to be borne out by the fact that many species of animals survived the Flood and continued to breed thereafter. The larger species, however, became extinct. Was this because they were too large and too dangerous to be rescued without endangering the lives of those on board and the safety of the ships themselves?

Moreover, perhaps humanity was, then as now, concentrated largely in huge urban areas. These large animals lived in great herds in areas remote from the centers of population. Herds of mammoth, for instance, would probably not live near large cities. Even if there were vessels large enough to take them, there would not have been time to mount a rescue operation, especially as these large and dangerous creatures would have been difficult to trap and load aboard a ship.

We know that when damming projects were undertaken in Africa in recent years, with the consequent flooding of thousands of square miles of territory, efforts were made to capture animals and reptiles, particularly species which would otherwise face extinction, and resettle them in areas elsewhere. If the whole world were to be subject to vast flooding, how much more important—and difficult—would such a project become, not only for conservation purposes, but also from a human survival point of view.

The theory of uniformism, according to which all geological and climatic changes have come about gradually over thousands of years, cannot possibly explain the deaths of millions of animals—and people—and the annihilation of entire species. Nor does it take into account the fact that subtropical flora existed not tens of thousands, but only thousands, of years ago both in Siberia and in other present Arctic regions.

Human experience, on the other hand, expressed through folklore and legend, has recorded catastrophic events that are borne out by facts.

The *Zend-Avesta*, holy book of the Persians, for example, states that Ormuzd, the Good Deity, gave the Aryans sixteen countries, described as a region of delight, as their home. Ahriman, the Evil One, turned their home into a land of death and cold, partly by means of a great flood.

Does this not remind us of the polar regions, which show evidence of warm climate conditions in the remote past?

It may be unfortunate that most of these traditional "memories" have been woven into the fabric of religious texts; religion and superstition have rendered less believable even the facts buried in these accounts. It may make more sense not to argue with the historical truths contained in a religious text like the Bible, but rather with interpretations of these events by those who later compiled and incorporated the facts within a religion.

By the time the collection of books known as the Old Testament was written in its present form (the fifth century B.C.), the *real* catastrophic events had not only been forgotten, but had become entangled with later, more localized disasters (like the Thera eruption), themselves attributed to the machinations of gods.

We assume that the Flood was an actual event with worldwide repercussions. We have theorized that the flooding was caused by a shift in the earth's orbital position. This shift in turn caused a sharp drop in temperature, leading to a rapid condensation of a great deal of water vapor suspended in the upper atmosphere.

The great drop in temperature must have been almost instantaneous, judging by the sudden freezing of the mammoths, although the course run by other aspects of the catastrophe would have been much slower. The flooding would have been exacerbated, caused by tidal waves of unprecedented proportions, earthquakes, and hurricane-force winds in many regions of the world.

Such violent events, leading to a drastic drop in temperature and subsequent flooding, must have had a triggering cause involving stupendous forces. Catastrophism theorists usually advance the concept of a *cosmic natural catastrophe*, which does have much to recommend it from some aspects.

The fact that the ancients apparently had time to prepare for the disaster, coupled with their astronomical knowledge, makes the celestial-encounter hypothesis an attractive one. They could have calculated the velocity of approach, the direction and angle, and the time/distance factor, to ascertain the period of maximum danger. From a study of celestial mechanics, however, we have seen that this kind of encounter would probably have destroyed the earth completely, or at least rendered it completely lifeless.

If it was not a natural disaster, was it an intelligently directed one?

The ability to cause widespread havoc on earth, indeed, to destroy life completely, is now within our power. Nuclear weapons, and some kinds of biological warfare, could kill all life on this planet. Existing forces whose secrets we have not yet learned are also capable of bringing ruin to the earth.

Prior to the test explosion of the first nuclear bomb at Los Alamos, New Mexico, in 1945, the scientists involved in the development of the A-bomb (the Manhattan Project) theorized that one of three things would happen at the point of detonation:

1. The device would not work at all;

2. It might initiate an irreversible nuclear "chain reaction" that would destroy the entire planet; or

3. It would explode in accordance with their mathematical calculations.

We are fortunate, perhaps, that it did work according to their calculations, although it may have been better if it had not worked at all. That these scientists conceived the *possibility* of universal destruction is a warning of what could still happen if nuclear reactions of uncontrollable power were initiated.

We are presently at the threshold of a dilemma: either we learn to control both the energies we have unleashed and our own violent natures or we shall one day destroy ourselves. Perhaps other races, elsewhere in space, have already made the wrong choice. Perhaps, at a time in earth's past, we came close to universal destruction.

A successful science-fiction film, *The Day the Earth Caught Fire,* treated the theme of the simultaneous explosion by the great powers of a powerful nuclear weapon which had the effect of moving earth from its orbit into a path of destruction toward the sun. Although much may depend on the exact locations of the explosions, it is doubtful that such a thing could occur in fact. The explosion of Krakatoa, equivalent to over 400 H-bombs, did not disturb earth's orbital path.

However, we would not need to go to such lengths to cause such havoc. Nuclear explosions centered in Antarctica could melt the ice cap, raise the level of the oceans by several hundred feet, and flood all major seaboard cities.

It has been suggested in some quarters that Superior Communities in space may have mastered the secrets of harnessing gravitational forces. Very little is known about gravitation except that, curiously, it seems to employ very little actual, measurable energy, but has tremendously powerful effects. Consider, for example, the strength of earth's gravitational field and the enormous amount of energy needed by spacecraft to escape from it. Perhaps gravitation, like electricity, is a form of energy—one we have not yet been able to detect because we lack the necessary knowledge and instrumentation.

A great deal of research is being done on magnetism and gravity, for obvious reasons. The ability to harness gravitic forces would be a major breakthrough, as great as the discovery of electricity or nuclear power. It would enable us to do away with the cumbersome, dangerous rockets with their vast amount of volatile fuels needed to escape the earth's gravitational field, making space travel both safer and cheaper.

Whether a "gravity engine" would be as useful as a high-velocity propulsion system for interplanetary or interstellar travel is uncertain. Although it would be useful if this were so, it would not be vital, since both the ionic and photon drives generate velocities approaching that of light. Either one of these sources of power could be put into operation once the "gravity engine" had released the vehicle from the gravitational pull of the earth.

A gravity-control device would have many other applications, particularly in the movement of materials. Masses of materials weighing hundreds of tons could be moved—"floated"—by neutralizing the effect of gravity. By means of levitation, masses of all kinds could be moved great distances without complicated transport systems or vehicles. Prefabricated parts of a dam, for example, could be levitated, fitted with small steering jets, and moved into any desired location by just one or two men.

On a larger scale, such a device could be used to alter the position of planets, perhaps by moving them into more favorable positions with regard to their suns. A planet orbiting at a distance too great for life to exist in comfort, although perhaps suitable in other respects such as size, mass, and composition, could be moved nearer to the primary, and made habitable. It has even been suggested that such a control, coupled with other advances in science, could be used to move earth through space to take up an orbit around another sun when our sun reaches a critical stage.

This sort of scientific speculation, once considered pure science fiction, is now indulged in by serious scientists. At the present time, in fact, scientists are experimenting with the use of powerful electromagnets at extremely high temperatures to suspend metals within a powerful magnetic field, known as a "magnetic bottle." Refinement of this technique on a larger scale could make it possible to suspend—and move—larger masses of material. Eventually, large units could be handled easily by making them virtually weightless.

At the same time, gravity-control devices could become weapons of fearsome destructive power. Tractor beams, which could immobilize electrical motors of any kind, and "gravity snatchers," which could wreak havoc by uprooting buildings and cities, or causing earthquakes and tidal waves, have all been themes of speculative fiction. Taken to the limits of destructive capability, such devices could be used to wrench a planet from its established orbit, or plunge it into a path of destruction into the sun, thus destroying the planet and its life.

Such things are acknowledged possibilities. Some scientists have speculated that such devices may already be in use by Superior Communities elsewhere in space. *Were such devices known in the remote past? Were they used in a great conflict, a conflict involving the earth, which resulted in a change in the planet's orbit, which in turn caused the lengthened year, the Flood, and its attendant destruction?*

Consider, for example, the curious biblical reference to the God who came with the weapons of his indignation, and who would "move the earth out of her place." (Isaiah 13:13)

The idea of levitation—of an ability to neutralize the iron grip of the earth's gravity—is as old as that of the philosopher's stone and the magician's wand, and could be rooted in sciences from an antiquity of which we have no knowledge.

Levitation devices may have been used either accidentally or deliberately, with catastrophic results. Perhaps they were used to shift slightly the orbital position of earth, and something went wrong. Perhaps they were used during a terrestrial conflict, or a conflict involving intelligent forces from beyond earth. If extraterrestrial forces were involved, there may have been an interval between some prior warning and its execution. This would account for the fact that there was a period of time in which to prepare.

Even if a weapon or device of this kind was used against earth by extraterrestrial agencies, it would seem that these weapons might have been known to earth's science in the past (we have the legend of levitation). This ancient knowledge would explain some otherwise puzzling pieces of evidence on earth.

For example, scattered across many parts of the world, in Europe, the Pacific, South America, and the Middle East, are the remains of what is called "megalithic architecture":

• Stonehenge is one example.

• The alignments of Carnac in France are another.

• In Britain, Ireland, and parts of France, there are the "dolmens," consisting of several uprights, often in a triangle, capped by a single huge slab, which, in some instances, weigh three hundred tons or more.

• The walls of Sacsahuaman above Cuzco in Peru are composed of blocks which weigh between seventy and two hundred tons each, and are so finely interlocked that not even a razor blade can be inserted between the stones. These walls have withstood the severest earthquakes.

• Megalithic underground chambers in Britain and France contain roofbeams weighing over one hundred tons.

• The Great Pyramid of Cheops is composed of 2.6 million blocks, each weighing between two and a half and ten tons, with slabs lining passageways weighing over fifty tons, and eighty-four weight-relieving beams, weighing eighty-seven tons each, placed one above the other over the main chamber.

In some places these megaliths exist in great numbers. What is more extraordinary, the stones of which they are built have been brought great distances. The blue stones of Stonehenge, for example, were brought to Salisbury Plain from Wales, over a hundred miles away. The stones at Sacsahuaman, Peru, were brought from quarries over twenty miles away, across a river and up to the top of the mountain. The vast slabs for the city of Tiahuanaco in Bolivia were brought either from the other side of Lake Titicaca or from islands within the lake.

In southern Brittany there is a single monolith called Locmaria-quer, atop a hill, which has now fallen and broken in three pieces. In its original state, the monolith was sixty-five feet tall and weighed over three hundred tons. It was brought from miles away. Even with our present lifting and transportation machines, we would have great difficulty moving such a stone. This massive monolith apparently formed only a small part of a vast megalithic system covering hundreds of square miles whose purpose is not yet understood, but appears, like Stonehenge, to have functioned as an astronomical computer.

Explanations of megalithic structures abound in contradictions. It has been said, for example, that the construction of the Great Pyramid employed 250,000 men for twenty years. (Recently revisions place the figure at around 2,500 men—which makes its construction by sheer brute force even less likely.) Yet experts on population in ancient Egypt say that the population of the whole country in the early period during which the pyramid was built was only about a million.

Out of a total population—men, women, children—of a million, it is doubtful if there would be 250,000 able adult males in any case. Even if there were, they could not all be employed on the construction of the pyramid. That would have meant that the entire male population of the country was employed at one site; as a result, there would have been no one left to function as soldiers, priests, farmers, administrators, scribes, and overseers, and in a host of other occupations.

With regard to Britain and Western Europe, the situation is even more complicated. Thousands of men and millions of man-days are presumed to have been spent in dragging and erecting the megaliths. Yet the population in these areas at the time of the megaliths' construction is reckoned to have numbered not in the thousands, but merely in the hundreds.

If thousands of people were involved, which suggests a large, stable, and highly organized population, where are the traces of the towns they occupied? The theory that the work force at Stonehenge, for example, was five thousand men, and that it took many years to build, implies a population of at least twenty thousand—the population of a medium-sized country town—living in the vicinity of the site. Even after five or six thousand years, such a population would be bound to have left fairly extensive traces, if only its rubbish, bits of broken pottery, etc. *There is not a trace* of a concentration of population anywhere near Stonehenge. No trace of any towns has been found—only the monuments, the great mounds, the rock-built "tombs."

Might the technique of levitation have been used to build these megaliths?

There is a tradition that appears in the mythology of the Americas that the priests "made the stones light," so that they were moved easily. Again, this connects with the legend of levitation, which may have referred originally to an actual technique or device, long since forgotten.

We have assumed that the catastrophe was caused by a great conflict in the past. The theory that a nuclear holocaust may have occurred on earth in the remote past is not new. Even if such a holocaust alone did not cause the alteration of the planet's orbit, it would still indicate advanced technology in more than one field of science.

We refer once again to the *Mahabharata,* which describes a weapon known as the "Brahma Weapon":

> When the weapon was discharged, smoke like 10,000 suns blazed up in splendor . . . then a thick gloom suddenly encompassed the hosts.
>
> All points of the compass were suddenly enveloped in darkness. Clouds roared into the higher air . . . showering blood . . . the world, scorched by the heat of that weapon, seemed to be in a fever.

Darkness hid the entire army. Then we beheld a wondrous sight; burned by the power of that weapon, the forms of the slain could not even be distinguished.

We can only infer from this that there was some sort of explosion. Reporting another such event, the *Mahabharata* goes on:

For many days there were terrible gales and people's hair and fingernails dropped out.

Food went bad and birds that had been ... [contaminated?] turned white, and their legs blistered and turned scarlet.

The elephants made a fearful trumpeting and sank dead to the ground over a vast area.

Then, for several years after, the sun and the stars and the sky were hidden by clouds and violent storms. It seemed that the end of the world had come.

According to the text, warriors who were in the vicinity of such a weapon had to jump into the water and strip all the metal from their bodies. They also had to wash themselves and all their clothing thoroughly, or the death-dealing breath of this weapon would have killed them. This death was lingering: hair and nails dropped out, the victims became pale and weak and eventually died.

The *Mahabharata* also mentions another weapon, Kapilla's Glance, which could burn 50,000 men to ashes within a few seconds.

We remember the same text's reference to fire:

"Then men tampered with the 'Divine Fire,' the earth split open and sixty million people in great cities were drowned in one terrible night."

In light of modern knowledge, it would seem that *this ancient manuscript, of unknown antiquity, describes the unleashing of nuclear weapons.*

The description of the explosion matches that of a nuclear blast and its destructive powers. The pattern of sickness that followed—"hair and fingernails fell out, sickness and weakness"—*the symptoms of radiation sickness.* They relate to no other known type of malady. How could people who lived thousands of years ago write so accurately about the effects of radiation *unless it had actually been experienced?*

Fifty years ago, the descriptions contained in this manuscript were not taken at face value, simply because no sickness of such a description was known fifty years ago. With the devastations of Hiroshima and Nagasaki, "myth" became reality. Moreover, the precaution taken by the people to prevent contamination and subsequent sickness—thorough washing in water—is now a standard precautionary measure for those who work with radioactive substances. Personnel employed in the vicinity of nuclear reactors, especially those who are suspected of any contamination, immediately shower thoroughly to wash away any radioactive particles.

Did the widespread practice of ritual cleansing by water—common to religions in many parts of the world—originate in these precautions in the remote past? The religious ritual was meant to cleanse from "sin"; it has no strictly hygienic applications. "Sin" and sickness, however, were synonymous in the past. Until the advent of bacteriology, disease was thought to be caused by wickedness or devils. It is possible that what was once a standard precaution against radioactive contamination passed, in the course of time, into religious rituals built around legends of the past.

Other interpretations of the events described in the *Mahabharata* and other religious texts seem incomplete, inconsistent, or not fully informed.

Erich von Däniken, in his book *Chariots of the Gods?*, assumes that the nuclear explosions were the work of aliens from space who interfered in a primitive terrestrial conflict in India. Perhaps, he reasoned, it was done to impress the natives, or to teach them some sort of lesson.

It is hardly likely, however, that technologically advanced visitors from other worlds, visiting a society of primitives armed only with spears, swords, bows, and arrows, would find it necessary to go to all the trouble of destroying cities with nuclear weapons in order to impress such natives. A few minor displays of technical "magic" would probably have sufficed.

Von Däniken applies this same reasoning to the biblical account of the catastrophic destruction of the "cities of the plain," Sodom and Gomorrah. We remember that Lot was told to hide in the hills and not to look back or he would lose his sight or his life. Lot's wife disobeyed and was turned into a pillar of salt. *Was she turned instantly to ash?* Afterward, Lot and his daughters hid themselves in a cave and thought that they were the last people left alive in the world.

Many scholars (notably science writer Agrest of the Soviet Union) have pointed out that the description (in ancient Hebrew scrolls not included in the Old Testament text) of the precautions to be taken closely resemble those given today to people in the vicinity of a nuclear weapon test.

Other writers have suggested that Sodom and Gomorrah were destroyed when aliens vaporized surplus stocks of nuclear fuel. Still others, with less exotic imaginations, have assumed that the cities were destroyed in an earthquake or volcanic eruption. Since the exact location of the cities is not known, geological tests cannot be carried out to prove or disprove these theories.

In any case, the account of Sodom and Gomorrah has elements in common not only with the *Mahabharata;* it seems also to have a connection with descriptions of the general destruction in the Genesis Flood story, as well as other legends relating to a great catastrophe. Since the episode of Sodom and Gomorrah refers to the time of wickedness, when the Lord was about to destroy the human race, it would seem to be more closely related to the Flood than to any later event.

The Old Testament account describes Sodom and Gomorrah as the "cities of the plain," without specifying any particular location. This is somewhat curious, as the Bible is usually at pains to specify exact places and names in common usage in the Middle East at the time the Bible was written. Yet from the description "cities of the plain," they may have been situated anywhere, not necessarily even in the Middle East.

The same holds true for the events described in the *Mahabharata,* especially the passage which states that 60 million people were drowned when great cities were destroyed in one night. The fact that the people were drowned when the earth split open, which finds a parallel in the Genesis account of the springs of the deep breaking through a great abyss, suggests a connection with the catastrophic events of the Flood.

At any rate, neither in the Middle East nor in India do we find traces of such cities as are described in the documents. If the cities *were* destroyed by nuclear weapons, they may have been completely vaporized, so that no traces would remain, anyway. More important, we find no evidence of geological upheavals of this nature in these regions. The destruction of Sodom and Gomorrah apparently refers to the same series of events described in the *Mahabharata.*

The events described in the *Mahabharata* were part of the heritage of the Aryan race, which invaded India thousands of years ago from the north, and whose written and spoken language has affinities with many Mediterranean languages, including Phoenician, Sumerian, and Etruscan. This language relation suggests a common point of origin for all these races. *Is it not possible that the Aryans brought these ancient legends with them,* that the legends were already in existence before the Aryans reached India? It would follow that the events described in these texts occurred at a time and in a place far removed from India.

The ancient text refers to the death of elephants: before the great extermination, there were over thirty species of elephant on earth, of which only two survive today. We have discovered no elephant graveyards in India that would suggest violent, large-scale destruction of the species. But we have proof that in the present Arctic regions, mammoths (elephants) and other animals died catastrophically by the millions. If this nuclear holocaust did not take place in India, then these elephants did not die there, either.

The biblical account of the "cities of the plain" may also refer to events that did not take place in the Middle East, but were a tradition brought with the people from other parts of the world. The Ten Tribes of Israel came originally from cities on the plains of Shinar, or Sumer (Mesopotamia). Abraham, the founder of the Hebrew nation, was once a citizen of Ur of the Sumerians, who before he left to search for a new God, worshiped the Sumerian gods El, Ea, and the moon god, Sin. Many Hebrew legends, traditions, laws, and customs stem directly from their Sumerian heritage.

It is interesting in this connection to refer once again to the Old Testament story of Noah and the Flood. It may have its roots in the Sumerian epic of Gilgamesh, which includes the Flood legend, and the story of Unapshitim, the Flood hero of the Sumerians. Furthermore, the name Noah is a corruption of an Egyptian name meaning "of long life," while the Sumerian name means "he of many years."

"Noah" may refer, then, not to a specific man, but to a group of people who had long life spans. These are the great ones of old, mentioned in the Bible—the same ones who escaped destruction in the Flood, and who had exceedingly long life spans. Further, the Noah of the Bible may have set forth in his Ark at a place far removed from the Middle East, although either he or his descendants settled there after the Flood.

Not only has it scarcely ever been stressed that the ancient legends may have had sources far removed from the culture centers where they were discovered, but little attempt has been made to give a rational explanation for the recurring movement of peoples through the ages. Until recent times, there has been an endless series of migrations, in a *north-to-south* direction. Surely this stems from the continual deterioration in climate, as the northern regions became progressively colder.

Consider, for example, the origin of the Sumerians. Although scholars have suggested that they may have migrated from more southerly or easterly regions, there is no proof of this. This assumption is based partly on the fact that the Sumerians built ziggurats, or step pyramids, which, this theory holds, served to remind the Sumerians of their mountainous homeland, perhaps in western India.

The Sumerians may well have hailed from more northerly latitudes, such as Northern Europe, where there are also mountainous regions. This is no more unlikely than an Eastern origin, particularly in view of mounting evidence that the present polar regions were subtropical thousands of years ago, prior to the catastrophe.

The strongest evidence for the existence of the Ice Age—which is also the strongest evidence of a great catastrophe—lies in the northernmost regions of the northern hemisphere. This does not mean that the catastrophe was limited in extent, only that its traces show more clearly in the large land areas of Britain, in the Arctic, and in Scandinavia.

It has also been suggested, from the evidence in legend, particularly in the *Mahabharata*, that this catastrophe could have been intelligently directed, either by accident or by design. Whether the conflict which led to the disaster was purely terrestrial or not, we do not know. If we assume that the ancient gods were terrestrial in origin, then the conflict was terrestrial; if the gods, or some of them, were extraterrestrial, then the conflict and the consequent destruction may have involved elements from beyond earth.

It is possible that the human race had an extraterrestrial origin,

and it could therefore be expected that a Superior Community elsewhere in space was either aware of the existence of intelligent life on
this planet, or that the human group here was responsible to such a
Superior Community for its actions.

Variations on the theme, most prominent in the Bible, that man
was to be punished for disobedience to God, recur in other mythologies: man disobeyed or aspired to rival the gods, and was therefore
punished. The "gods" in all these legends have two things in common:

• They always have an origin in the sky.
• They appear in human form, but with superhuman powers.

On the basis of these shared characteristics, Von Däniken, among
others, has assumed that these legends depict visits by superior cultures from elsewhere in space, that the "gods" were in fact astronauts,
and that they exercised temporary control over earth and its inhabitants.

Were they gods or not? The question is not easily resolved,
especially when we are guided solely by mythologies and legends
largely incorporated now into religious institutions, and when so
much depends on individual interpretation.

The story of Exodus, in which plagues are attributed to God, or
the actions of God, seems to describe the results of a disastrous natural
phenomenon. Even today such natural disasters are termed "acts of
God." Mythologies stemming from an earlier period, however, dealing
not with the actions of gods, but with the gods themselves, seem to
describe intelligent entities—actual recognizable people, with superhuman powers and abilities, who came down from the sky. This is no
less true of the Old Testament than of any other religious writing;
Yahweh of the Old Testament has much in common with Zeus, Odin,
and Osiris. In fact, the Old Testament reflects a curious dualism: God
is not only an invisible, omnipresent *spirit*, a force which animates all
things; God is also an actual *physical* presence with superhuman
powers.

According to Exodus 19:16–22, Moses went up onto the
Mount, where he saw the feet of God, standing on a pavement made
of crystal. In other words, he saw the feet of a person standing on a
glasslike platform. Now if this was not mere imagination or fantasy
—and it is difficult to imagine that everything of this nature in the
Bible is—then the feet were attached to legs, and the legs to a body,
and the body to a head. Although Moses never saw the head—he was

told not to look, and no doubt his superstitious fear was such that he would not—no doubt the head was human, although it may have been hidden by a mask or helmet of some sort.

A "god" who appears to be solid and of human form is not the same thing as a celestial power who occupies all space and all time; he is rather an actual physical object and as such has physical limitations. No matter that such an entity has an enormously extended life span and may command stupendous powers and control natural forces beyond those we control: this is still an *intelligent entity not beyond our present comprehension*, although it may have surpassed the understanding of people in Moses' day.

The description of Yahweh, including the oblique references to his great age, seems strikingly like the description of the people who lived before the Flood, who were also clever enough to rival the "gods," and who lived for great periods of time.

Before Moses went up to the mount an odd-shaped cloud descended amid flames and thunder, and a great voice told the people to stand back lest they be killed. This seems a curious way for a "god" to arrive, and the warning is also peculiar. Read in terms of modern experience, could this not describe the landing of a spacecraft of some kind? Flames and thunder would seem to describe—fairly accurately, at that—the action of a jet braking mechanism, which would obviously be dangerous and even fatal to approach. The great warning voice is no miracle, merely an amplified broadcast from within the vessel.

We can look at all these things from two points of view: either they were hallucinations of some sort, or they were *real events and happenings which were not understood by those who witnessed them.*

Many events described in the Bible—and elsewhere in ancient writings—correspond too closely to modern reported sightings of UFOs to ignore this connection. The description Ezekiel gives at the beginning of his narrative, for example, reads extraordinarily like a modern report of a UFO sighting, even to the nature of movement and the colors associated with them.

The pharaoh Thutmosis III saw "fire circles" moving in formation across the sky, and observed that they gave off a stench. His priests were unable to tell what they were, and apparently they drew no conclusions. However, it is of interest to note that some modern observers of UFOs report that they glow (fire?) and that their passage leaves a highly pungent odor. The action of the UFOs' drive mechanism could cause ionization of the atmosphere through which they passed, which would not only create the color effects described by Ezekiel and others but would also cause a concentration of pungent-smelling ozone.

Responsible authorities generally discount any suggestion that there may be extraterrestrial vehicles in our atmosphere; even the most hidebound among them admit that 5 per cent of these sightings cannot be explained—or explained away. It is this 5 per cent which seems so closely related to ancient writings; if we assume that in the instance of this 5 per cent, there may be machines from other worlds investigating us, then similar reports in the Bible and elsewhere may have the same significance. We must also take into account that the world then was in a fairly primitive state, lacking sophisticated weapons and electronic detection devices, which would make it easier and safer for any extraterrestrial society to explore the earth, both from the sky and the ground, at will.

However, the connections may be even closer. Many ancient narratives contain the theme of the anger of the gods in the face of man's defiance of them. In the Bible, as in other religious texts, man's defiance was represented by the fact that man held too much knowledge. Man ate of the "Tree of Knowledge"—and sinned. Apparently there were certain unspecified areas of knowledge that should have been closed to him.

There could be many reasons for the "disobedience" and hence for punishment. If the gods who placed man on earth and taught him the rudiments of civilization were members of a Superior Community, it is possible that they gave strict instructions on the paths this civilization was to follow. Or perhaps they were to follow certain instructions regarding their colony, or perhaps they discovered things and had intentions which would have made them a menace to a Superior Community elsewhere in space.

If the earth colony posed a serious threat to others—it is said that man challenged the supremacy of the gods—then the action the gods took was designed to eliminate man as a serious rival. Following the destruction of most of humanity and its civilization, many millennia would pass before earth's colonists were in a position to challenge any outside forces. Also, a subsequent propaganda campaign directed by the Superior Community against the now primitive survivors, by playing on their newly developed superstitious fear of the alien "gods," would further retard progress. A study of religions up to the present day surely supports such a view.

One can also judge that man may not, in the beginning, have meekly surrendered to his fate. This may have resulted in the "war of the gods," fought between the mighty ancients and the "gods" from space, both of whom became "gods" in the minds of later generations. We have a hint of the kind of weapons used in this conflict, from a reading of the *Mahabharata:* nuclear, possibly lasers, and gravitic devices.

Is the foregoing merely science fiction dressed as a hypothesis? Anyone may be forgiven for thinking so. But let us remember that some passages in the Bible might also be considered "science fiction" and therefore also dismissed.

Either there is a core of truth in all or there is no truth at all. We have suggested that the God of the Old Testament did actually exist—as a real, intelligent entity. We have also shown that many events describe *actual happenings,* like the Thera eruption, about which little or nothing was known until recently. Further, we have shown that the Flood described in the Bible, as in other manuscripts from the remote past, was a catastrophic event for which there exists geological proof.

Can we find a further clue in the Bible about the events leading to the catastrophe? The book of Isaiah contains an extraordinary reference, which has no counterpart elsewhere in literature:

Isaiah 13:3-5: "I have commanded my sanctified ones, I have also called my mighty ones for mine anger, even them that rejoice in my highness.

"The noise of a multitude in the mountains, like as of a great people; a tumultuous noise of the kingdoms of nations gathered together; the Lord of Hosts mustereth the host of the battle.

"They come from a far country, from the end of heaven, even the Lord, and the weapons of his indignation, to destroy the whole land."

In verse 13: "Therefore I will shake the heavens, and the earth shall move out of her place, in the wrath of the Lord of Hosts, and in the day of his fierce anger."

And in chapter 24, verse 1: "Behold, the Lord maketh the earth empty, and maketh it waste, and turneth it upside down, and scattereth abroad the inhabitants thereof."

No statement of this nature was made in relation to the Exodus episode. Is it not possible that this, like the narrative of Sodom and Gomorrah, related to the Flood?

"The earth shall move out of her place," Isaiah observes, "and be turned upside down." Does this refer to the time when the earth was moved out of its orbit, causing the Flood?

Verses 3-5 in chapter 13 do not say that the anger of the Lord of Hosts was manifested in a natural catastrophe. Do these verses refer to forces directed from *beyond the earth?* The ancients were familiar with astronomical facts—the Bible states (Job 26:7) that earth and the moon "hang upon nothing," so that we can reasonably infer that they knew the earth hung in space, and was not supported, say, on the backs of elephants. Had a natural catastrophe occurred—like the collision of a comet—surely it would have been mentioned as such.

This visitation of a great catastrophe, then, does not seem to originate in natural causes; it seems to have been engendered by some intelligent entity. But surely this intelligent force is not the same God who created the heaven and the earth, and all the stars, and who breathed life into all living things? An all-powerful, invisible god would not have needed a host of followers, or weapons of any kind; he would be able to accomplish anything unaided.

What we have here is obviously a person, or a group of people.

The Lord and his host, the text goes on, "came from a far country, from the end of heaven," Translated into modern terminology, it would seem that the "far country" in heaven refers to another world in the sky—*another planet in space.*

At the time such a statement would have been recorded, it would probably have had no literal reference point; by then, astronomical science would have been forgotten, and the stars thought of as mere points of light. The reference to another planet in space could only be understood by such people as another country in heaven; even this might not have made much sense to them.

The belief, common to many religions, that the stars were the abode of the gods was probably not understood by later generations, but it is significant that the stars were thought of as the abode of the gods.

Let us return to the theme of the connection between the gods from the sky and the people of the earth. We have seen that there is a great deal of resemblance between the gods described in ancient literatures and the race described as living before the Flood. The terrestrial "gods," *and* the gods who apparently lived among the stars, *and* the race of people who lived before the Flood, were all of the same origin.

This view is supported by statements in Genesis, where we are told that the sons of God mated with the women of earth and produced children (this, in the time *before* the Flood). The mating of gods from the sky with women on earth is a theme common to many mythologies. Presumably, then, they must have had a common origin.

Was "God" originally of a primitive race, which later gained great knowledge and began to travel through space? This statement does not seem to define some vague spiritual kinship; it seems to portray a more concrete reality, especially if the "God" in question is an actual personage, not an all-pervading spirit. Many of the gods of antiquity were deified humans.

It appears that by the time the early civilizations rose, communications between the "gods" and man had ceased. Certain channels of communication may no longer have been available to earth, either because of the destruction of the high civilization and its means of communication or because those channels of communication were withdrawn. (We have previously suggested that interstellar channels of communication exist which we have neither the knowledge nor the technology to detect.)

One thing is certain, however: *there were survivors of the catastrophe*. Their accounts of the events, however, have by now been distorted, and have passed exclusively into religious mythology. The original truths have all but vanished. These survivors also passed on, in less distorted form, their knowledge of all those things essential to the rebuilding of civilization.

The areas that felt the major impact of the catastrophe are the present Arctic lands—Siberia, Scandinavia, and Alaska—the areas now beneath the Arctic and North Seas, and to a lesser extent Britain and parts of Western Europe. These areas may have been primary zones of the high civilization prior to the catastrophe, although such a culture was probably worldwide.

We do not know what may lie beneath the ice of Antarctica. If it is true that this was a habitable continent only a few thousand years ago, traces of civilization must lie beneath the thousands of feet of ice.

At Ipiutak on Point Hope in Alaska, archeologists have found the remains of a large town, of which only slight traces now remain: the outlines of long, broad avenues in a regular pattern. The town, which does not resemble present-day Eskimo culture at all, is thought to have been large enough to support a population of many thousands. The artifacts discovered in the area suggest that this was far from a hunting community.

How was such a large community able to survive in a region that is now permafrost? Today the area can support only small bands of hunters who do not live in permanent settlements. The existence of towns like Ipiutak—and there may be others, buried beneath the ice—makes sense only if the climate was once totally different.

So far, nothing has been found that would indicate that this was a highly advanced society. Since the only artifacts found were associated with burials, however, this proves very little. Archeologists thousands of years hence may draw similar conclusions about *our* civilization if all they have to go on are the contents of a cemetery.

The traces of Ipiutak are slight: we have found only the vague outlines of streets, and can have no idea what their buildings may have looked like. If they had possessed any kind of machinery, it would have rotted and rusted away within a few centuries. But how long would our abandoned machines last, if our civilization were destroyed? A modern automobile, for example, would rust away within a few years; after a century or so, scarcely a trace would remain—and after several thousand years, not even the dust would be left.

In much the same way, although a great deal has been written about so-called Neolithic man, we do not really know whether he was primitive or not, according even to our standards. There are, to be sure, no traces of the existence of a high civilization; but there are anomalies not easily explained if we assume that Neolithic man was totally without civilization:

• These cave-dwelling ancestors of ours had short hair and clean-shaven faces, clothing that resembles that of modern man, including evidence of buttons and fine needles.

• Their drawings resemble those done today more than they resemble those done in any other period of history. A group of etched figure studies from Neolithic times look like jottings from a modern artist's notebook.

These two aspects—the appearance of man and his art—show us a people more like us than many of our more recent ancestors. Significantly, *both the art forms and the artifacts show a deterioration or degeneration* in later Neolithic times (nearer to the present), surely the reverse of what the situation should be. As man traveled further along the road to intelligence and manual skills—as proponents of the evolutionary concept hold—the art forms and the skills should have improved.

What if we take a totally opposite point of view? What if these cave-dwellers were not people emerging from the primitive, but survivors of a highly advanced civilization who were forced to live a primitive existence?

Science Year (1966) reported this item in connection with an anthropological conference held that year: "Many of the so-called primitive peoples of the world today, most of the participants agreed, may not be so primitive after all. They suggested that certain hunting tribes in Africa, Central India, South America and the western Pacific are not relics of the Stone Age, as had been previously thought, but instead are the 'wreckage' of more highly developed societies forced, through various circumstances, to lead a much simpler, less developed life."

Let us look again at this problem from the other end of the telescope. Let us assume that our civilization were destroyed in a catastrophe of some sort—a worldwide natural disaster or nuclear war. Those who survived the destruction of the cities and cultivated areas would flee from the ruins to seek wilder country where food could be obtained.

For the inhabitants of our complex urban civilization, thrust suddenly into a state of nature, survival would be extremely difficult. Most twentieth-century people have very little idea how to look after themselves. The ordinary office or factory worker shops in a supermarket or department store, uses appliances and means of transportation about which he knows very little. How many people would have any idea of how to make clothes, trap or hunt animals, or make pottery to cook and store food? How many, without the aid of matches, would even be able to make a fire?

These questions may seem pointless, until we realize that modern man has been so cushioned by the civilization we have created that he would be virtually helpless in a savage world. Those who survived would be those who had painfully learned to cope with their new environment. (Recall that military units undergo training courses in survival in the wild.)

The traces of cave-dwellers of antiquity manifest a blend of sophistication and primitiveness that is curious indeed—unless we take the view that *they were once a civilized people*. Sometimes we find evidence of fairly high quality pottery, for example; some, however, is extremely crude, and in some places we find none. If, as the evidence suggests, these were all the same kind of people, living under roughly the same conditions, at roughly the same time, could we not conclude that some were coping better with their environment than others? Perhaps some of the groups had someone who knew about pottery-making.

We have found cave paintings of animals and, less often, of people, which are extremely well drawn, etched and painted under conditions that even modern man would find almost impossible. Does this not suggest that these artists were in fact descendants of civilized and artistic people, who used techniques with which we are not familiar, and painted for no other reason than that it was something to do? In this context, perhaps our theories about hunting rituals and sympathetic magic need another look.

These cave-dwellers were not part of a large, permanent population; they lived in small groups, under difficult conditions, on a temporary basis, some with good weapons and utensils, others with poor weapons and equipment. This state of affairs could be expected in the case of a highly civilized people thrust back into a state of barbarism, deprived of both their civilization and a means of communication with fellow survivors.

Furthermore, why should these people, if they were truly primitive, appear to have spent so much of their time carefully painting accurate representations of animals on the walls and ceilings of their caves? Given the life-style ascribed to these primitive peoples, even the few involved in this work would have been needed for the more important problem of staying alive. Going partly on the observations of present-day primitive tribes, with their hunting ritual and ceremonies, anthropologists theorize that these cave paintings may have been executed in connection with some magical hunting rites. There is, however, a vast difference between present-day primitive peoples and these traces of ancient man, and no comparable paintings exist to show a similarity between the two.

Curiously, modern scientists, especially anthropologists and archeologists, appear eager to ascribe a magical or religious significance to almost every activity of ancient man. Generally speaking, there are few traces of purely religious objects in association with Neolithic man. The small sculptures that have been found—usually of the female form—may not have had any religious significance at all. A finely done head of a young girl, carved in ivory and only an inch and a half high, was found in Landes, France. Thousands of years old, it shows an elaborate hairstyle, seems neither of a religious nature nor the depiction of a member of a totally primitive community.

Remains of humanity from prior to and immediately after the catastrophe show little trace of any formal religion, which apparently came much later. It seems likely, in fact, that most religion has as its basis the events and the personalities of the era of catastrophism itself.

This does not mean that Neolithic man was not sufficiently advanced to have developed a religion. Generally, the more advanced a society becomes, the less it seems to have to do with organized religion. The increase of knowledge and the consequent lessening of the influence of superstition deposes many gods. Perhaps Neolithic man, as the descendant of a high civilization, had largely dispensed with religion, which was "reinvented" as humanity reverted to a more truly primitive state.

The importance of catastrophism in the Hebrew—and consequently the Christian—religion lends support to this view. The same holds true for much of the religious writing of the Hindu religion, the Sumerian, and the pre-Columbian cultures of the Americas.

15 THE RE-EMERGENCE OF CIVILIZATION

> "No more surprising fact has been discovered by recent excavation, than the suddenness with which civilization appeared in the world. It was expected that the more ancient the period, the more primitive would excavators find it to be, until traces of civilization ceased altogether and aboriginal man appeared. Neither in Babylonia nor Egypt, the lands of the oldest known habitations of man, has this been the case."
>
> —P. J. WISEMAN, in *New Discoveries in Babylonia about Genesis*

Every civilization of antiquity—including ancient Egypt and Sumer, and their "descendants," Assyria and Babylonia—seems to have appeared fully formed:

• One of the oldest urban centers so far unearthed is Catal Huyuk on the Anatolian plain in Turkey, which dates back to the fifth and sixth millennia B.C. It is said that the grains and vegetables cultivated by the inhabitants were already well developed, and had been introduced into the area from elsewhere.

• The Maya civilization of Yucatán and other regions of Central America manifest no transitional stages. Their complex glyph writing, advanced mathematics, and elaborate stone cities appear to have arisen spontaneously, with no crude beginnings.

• In Peru and Bolivia, the Inca built their empire on existing cultures—cultures that seem to have no transitional stages. The city of Tiahuanaco, one of the most remarkable collections of ruins on the

160

face of the earth, is built on a bleak and barren plateau as if out of thin air.

In each case, the civilization was transplanted from elsewhere—or the knowledge to create the civilization originated elsewhere. For this reason, we hypothesize that a high culture once existed which, although destroyed, left behind a core of knowledge upon which later cultures were built.

Virtually the only clues we have about the origins of these cultures are legends of their foundation. These legends, however, are extremely instructive; furthermore, they lend a great deal of support to the hypothesis mentioned above.

According to many of the legends, the civilization was built by a creator god who usually arrived by sea. (Some legends describe gods who come down from the sky.) We shall discuss three of these creator-gods legends—those that have to do with the Cretan, the Maya, and the Sumerian civilizations. Interestingly, they seem closely connected with the Old Testament account of Noah, as well as other Flood heroes.

The Atlantis described by Plato—which we have already identified with the ancient Cretan sea empire—was, according to legend, founded by Poseidon. Poseidon, a Mediterranean sea god, was given the island of Atlantis and had children by a mortal woman. Poseidon arrived on the island from the *sea:* was he (like Noah, Gilgamesh of the Sumerians, Quetzalcoatl and Wira-Kocha of the Americas) a Flood survivor? Poseidon's designs for his new kingdom—canals, aqueducts, bridges, and buildings—indicate a type of knowledge that must have had a source, knowledge that was apparently brought to the island by one man or by a group.

As in the case of Noah, Poseidon (which may not have been his original name) was perhaps the leader of the group which bore his generic name. There are many examples of legends through which a real human being becomes a god after his death. In the course of time, it is forgotten that the god was once a man, and he therefore becomes a purely mythical figure. Mythical figures do not, however, build earthly empires.

As in the case of Egypt and Sumer, the Cretan Minoan civilization (Atlantis) had an abrupt beginning. The buildings excavated on Crete show an extreme sophistication of design in all their aspects. Reconstructions of buildings in model form, based partly upon the ruins and partly upon drawings, show us structures in no way inferior in design

or construction to those we build today. In fact, the ancient drainage
and water-supply systems would seem superior to our own.

> When Solon was in Egypt, the priests explained to him that the Greeks did
> not remember clearly either the destruction of Atlantis or what went before,
> because they had forgotten the art of writing. The priests of Sais implied
> that there had been a break in Greek civilization lasting many centuries. If
> this were so, the original ancestors from whom the Greeks were descended
> would have been the survivors of the precatastrophe civilization.

As is true of many other cultures, it would appear that the earlier
period of Minoan culture was far superior to the later phases—totally
the reverse of what should be expected if these civilizations were
emerging from a primitive state. Although the culture was catastrophically destroyed by the Thera eruption, it was already at war with
the emerging Greeks, who were at this time barbarians from the north.
The Cretans would no doubt have been absorbed eventually into the
more vigorous Greek world, which in turn was absorbed into the
Roman.

On the other side of the world, in the Yucatán region of Central
America, the Maya built their cities. As is the case with Old World
cultures, there does not seem to have been a period of transition from
savagery to civilization. The great, highly decorated cities appear
suddenly in the jungle. Extremely complicated glyph symbols found
carved on buildings, temples, stairways, and stelae have no traceable
early form. Some authorities have suggested that there may have been a
preliminary phase during which writing was developed on wood (or
paper), which has not survived the passage of time. This, of course, is
theory. Unfortunately, many aspects of the Maya culture are literally a
closed book, since we are able to translate only the mathematical
symbols, and these without any degree of certainty.

Much of the knowledge we have of the Maya stems from transla-
tions of Mayan texts undertaken by the Spanish friar Diego de Landa.
According to the legends contained in these texts, the first Maya cities
were built in the remote past by the Saiyam Uinicob—"the Old White
Fathers"—who were later destroyed in a flood. There seems to be a
connection between the Saiyam Uinicob and what was called the
Great Arrival, led by a god (a deified human?) named Itzamna, who

was followed some time later by another god, Kukulkan, reputed to have been a white man.

The people who are credited by the Maya with the building of the first cities were white men who arrived by sea. Actual representations of these ancient gods exist in Central America: stelae depict likenesses of bearded, Caucasoid-featured men, and wall paintings in Maya temples bear the likenesses of pale-skinned, blond-haired people.

These were not paintings of the white people who came to the shores of the Americas at the end of the fifteenth and the beginning of the sixteenth centuries; some of the sculptures and paintings date from before the Christian era. Further, the Maya civilization had collapsed prior to the tenth century A.D., long before the Spaniards first came to Middle America. (The Vikings are not thought to have traveled so far south; in any case, the Vikings of those days would hardly have been able to create the Maya civilization.)

Radiocarbon dating techniques have established that remains of the Maya civilization precede by thousands of years the arrival of any European travelers. Wooden lintels found on a building at Tikal in Guatemala, for example, were dated at approximately 1500 B.C., and Dzitbalchaltun ("where-there-is-writing-on-flat-stones") in Yucatán, has been tentatively dated at 2000 B.C. Older sites may yet be discovered, particularly in the uninhabited areas of the Quintana Roo. Tikal, Coba, Uxmal, and other Maya cities may be far older than has been thought.

Using the Maya mathematical symbols, the date of the Great Arrival, the start of the Maya culture, has been placed at 3373 B.C. We do not know whether this date is accurate; it has not been possible to establish a correlation between the precise Maya dates and our calendric year-dates. While we can determine the month and day for Maya dates, we have had to rely on radiocarbon dating techniques to estimate the year of the beginning of the Maya calendar. We have assumed that a calculation that spans a period of say, five hundred years, commenced in A.D. 100; this, of course, is speculative—the commencement could have been centuries earlier, or much later.

However, 3373 B.C. may not be too remote a date, although most traditional scholars believe that the American cultures are more recent than those of the Old World. Interestingly, the dates for Tikal (1500 B.C.) and Dzitbalchaltun (2000 B.C.) compare favorably with Old World cultures. Tentative calculations gained from the Maya calendars continue to push the dates back even further.

At any rate, the Maya legend describes the building of a civilization by highly advanced people who arrived by sea at a time and in a manner very much like the legendary founding of the Minoan sea empire by Poseidon: two cultures, on opposite sides of the world, both of which started abruptly at about the same time, and both created by seaborne gods. *Were Poseidon of the Cretans and Kukulkan of the Maya both Flood survivors from the high culture destroyed in the cataclysm?*

Traditionally, the advanced civilization of Sumer was founded by Gilgamesh, whose story, the Epic of Gilgamesh, parallels exactly the Old Testament story of Noah. Gilgamesh was described as part man, part god, and lived longer than ordinary mortals. Like the Maya and the Minoan, the Sumerian civilization was founded before 2000 B.C., introduced from some unknown location, fully formed.

Most authorities agree that the "culture/Flood heroes" were largely responsible for the creation of these ancient urban civilizations, and were venerated by later ages as gods. The movements of peoples into the areas where the civilizations arose, and the arrival of these culture heroes, are, it seems, scientific and legendary descriptions, respectively, of the same phenomenon. *Poseidon, Osiris, Itzamna, Kukulkan, Quetzalcoatl, and Wira-Kocha, now regarded as individual gods in mythology, may more properly refer to groups—perhaps large groups—of human Flood survivors.*

If such people, equipped with the knowledge in many fields necessary to re-create urban cultures, brought this knowledge with them, they must have come from a still older center, or centers, of civilization that we have not been able to locate.

In an effort to locate one particular source, many people have seized upon the Atlantis theory. If, however, civilization prior to the Flood was worldwide, there may have been many centers of high civilization, concentrated in various places, as there are at the present time.

Let us assume that one of these great centers was situated in the northern hemisphere, in the area that now comprises the Arctic lands.

After the catastrophe, the survivors were forced to flee—perhaps by boat, perhaps by air; others may have survived the catastrophe in high places like mountaintops. (Remember that many of the gods of antiquity lived on mountaintops.)

These survivors would have fled southward, away from the center of disturbance and the area where the most drastic changes in climate occurred. They could have split into two main streams, one heading southeast, toward the Mediterranean and Middle Eastern regions, and the other, southwest, toward the Americas, through devastated North America, to find refuge in Central America and the Pacific coastal areas of South America.

Some of these survivors may have wandered for many years before they found a place to settle. They may have found, by the time they arrived, that those in the area who had survived the catastrophe—or earlier arrivals—had reverted to a primitive state. The collapse of a large, highly organized civilization can result in a relatively rapid decline to barbarism. After the collapse of the Roman Empire, for example, the last Roman legions left Britain around 420 A.D. By the seventh and eighth centuries, the reversion was almost complete. Towns were destroyed, and scarcely any knowledge remained of the high urban culture enjoyed by the Romanized inhabitants. This decline was caused by localized destruction, by a human agency with primitive weapons; if the whole world were affected, particularly if weapons of mass destruction were used, the results would clearly be much more drastic.

There may also have been a population drift northward from the Antarctic continent, warm prior to the catastrophe. The remains of civilization may be buried at the bottom of the Ross Sea or under thousands of feet of ice.

Although there are many similarities between the general cultural outlines and mythologies of ancient Mediterranean and ancient American civilizations, they differ markedly in detail. The contrasts—for example, the total differentiation in written languages (where they exist), and the complete failure of the New World cultures to make any practical use of the wheel—are difficult to reconcile.

Two explanations for these differences have been offered:

The differentiation could have been a later development, becoming more marked as time went on and the groups on either side of the Atlantic became more isolated from each other, until each forgot that the other existed. We note that the American cultures did not use the wheel; neither did the Egyptians in their early stages. The Sumerians used a form of pure picture writing that did not differ greatly in style and concept from that developed by the Aztecs. As the Egyptians later developed hieroglyphs, the Sumerians cuneiform, and the Chinese ideograms, the complex Maya symbols may have developed similarly. The fact that the vast Inca empire apparently did not use a written language did not deter them from complex and highly efficient administration.

Secondly, differentiation between Old and New World cultures may have depended on what kind of people arrived in different areas. Today we have experts and specialists in many fields, but outside their own specialties they are as ignorant as any layman.

Such a situation in the postcatastrophe world would explain the extreme sophistication in some fields of knowledge and almost complete ignorance in others. It may all have depended which of the survivors knew what.

Possession of certain foods or seeds, too, may have depended on what a given group was able to salvage when it fled the deluge. For instance, why was maize, which grows in many parts of the world, found originally only in the Americas? Why did the horse, which thousands of years ago appears to have roamed the American continent in great numbers, die out, to be reintroduced by the Spaniards in the sixteenth century?

Did the groups who traveled to South America take seeds of maize with them from the drowned lands of the north, while those who traveled to the Middle East did not? Is it possible that most of the horses of the Americas were destroyed in the disaster, while a few living elsewhere were saved? We remember the legends of the arks, carrying people *and animals*.

The survivors who made their way to the Americas found themselves in extremely difficult terrain, unlike the more easily traversed lands of the Mediterranean basin. The deterioration in climatic conditions was also more marked in the Americas. The groups were isolated from each other, and the isolation became more complete with the passage of time. This would explain why the Maya, Toltec, and Inca cultures were unaware of each other's existence.

Traces of an extremely archaic culture in the American continent suggest, however, a degree of homogeneity among all these cultures in the past—possibly at their point of origin. The mythologies of all the major American cultures whose legends we know refer to the Flood and the sea gods who created civilization and then left, promising to return at a later date. (The cult of the cat god is also widely distributed throughout the ancient American cultures.)

The origin of the Inca, who were the ruling class of the Peruvian empire, is highly obscure. Their mythology seems to link them with the mysterious white gods (the culture bearers). As such, they would have been direct descendants of the original civilizers, parallel to another original post-Flood civilization, that of Sumer. Remarkably, there appears to be a great deal of similarity between Quechua, the language of the Inca, and ancient Sumerian. Some of the architecture found in Nimrud in Assyria bears striking resemblance to that found in Tiahuanaco in Bolivia, reputed to have been built by the Wira-kocha, or culture bearers.

We have suggested that the survivors would flee from the area of greatest destruction in two main directions, and once settled, would re-create civilization—without the benefit of the sophisticated technology that had been destroyed.

Those who were unable or unwilling to flee, or who took refuge on mountains or in deep shelters, on the other hand, would become progressively more primitive as time passed, attempting to cope with the harsh new conditions. Eventually, as the climate became too severe, they would have to forsake their lands and look for more favorable conditions to the south.

Each wave of migrants traveling south would have lost more knowledge of civilization than the one before. They would eventually come into contact with the earlier rebuilt civilizations. The resulting clash between the new barbarians sweeping down from the north and the settled civilizations would bring about the destruction of the civilization or the substitution of a more primitive society.

In the case of the ancient cultures this is exactly what happened.

Stonehenge and the other great megalithic structures were built more than a thousand years before the time of the Druids (ca. 100 B.C. to A.D. 100), by which time both their purpose and the techniques for their construction had long been forgotten. Later peoples were unable to duplicate these great engineering feats. In fact, the megaliths were regarded with awe and were thought to have been built by gods. They were used as places of worship.

In the Middle Ages, superstition about these monuments reached such proportions that the church condemned them as the work of devils and symbols of pagan forces, and tried to destroy them. Some were taken away and destroyed, but only after they had been hacked to more manageable size. Muscle power—and even oxen power—was not enough. Even with today's machines, it has been found extremely difficult to replace fallen stones at the Stonehenge site. None has been restored as accurately as the original builders had placed them.

Megalithic building, of course, is found in many parts of the world. The pyramids were built before the growth of Egyptian civilization proper, and were as much a mystery to the ancient Egyptians as sites like Stonehenge are to us today. The Incas of Peru had no knowledge of the building or the builders of the great megalithic structures which abound in Peru and Bolivia.

Were these great monuments of a fabulous past the reason for the legends of the giants?

Let us imagine our primitive ancestors, centuries after the catastrophe—poor, ill clad, ill fed, short-lived, and prey to many diseases, wandering through cold desolate landscapes, searching for new and greener lands to settle. Scattered across the lands they traversed were the relics of a bygone age: giant ruins, the remains of broad highways and huge buildings.

Perhaps in those days there were many more such structures than there are today; some of them may since have been buried under desert sands, or beneath the sea, shattered and buried by earthquakes. (Of the Seven Wonders of antiquity, only the pyramids remain today, and they were built more recently than the relics of the antediluvian world.) Surely these early descendants of the Flood survivors, still haunted by the legends of past greatness and a terrible disaster, thought themselves to be in lands where giants once walked and lived. How, they would have thought, could mere man have raised such masses of stone? What man would need paths as long and as wide as a modern highway, in a day man traveled on horseback or on foot?

The ancient urban cultures, in their earliest phases superior to those which followed, cannot be reconciled with an emerging, evolving civilization. The megaliths and their builders, who were either contemporary with or preceded the early city builders, represent one of the world's great riddles. We shall never solve it as long as we continue to think of them as great but slow and agonizing accomplishments of primitive Stone Age peoples.

THE RIDDLE OF THE
GREAT STONES

Megalith means, simply, "great stone," from the Greek *mega,* large, and *lithos,* stone. A monolith is a single upright stone, a trilith is two uprights capped by a crossbeam.

Before the first post-Flood civilizations were created by the Cretans, the Egyptians, the Sumerians, the Olmecs, the Maya, and the forerunners of the Inca, megalithic structures were being erected in Europe. Some in Britain, and others in places as far apart as Peru and Egypt, may have been built before the Flood.

The great mysteries of how and why they were built are an integral part of the hypothesis developed in this book. Most previous theories about the megaliths have missed one or two very important points, which we will seek to elucidate.

The first peculiarity we observe is that they were all built *before the wheel was either known or put to any use.* One would have thought that megalithic building would have *followed* the introduction of the wheel, not preceded it.

Perhaps, however, the wheel was *not* man's first great technical invention, but was a subsidiary device used widely only after other, better, and more efficient methods had been lost. The wheel was not, in fact, a step forward, but a step backward.

In any case, the invention of the wheel or roller does not really solve the problem of how the megaliths were built. Many of the giant stones used in megalithic construction were not only brought great distances; in some instances, they were raised to considerable heights. The main problem is one of size and weight, and the unpowered wheel is of little or no use in this respect. Furthermore, we are assured by the experts that neither the wheel nor the block and pulley was used by any megalithic builder.

Legends of the existence of giants, we have seen, arose because people thought that ordinary men could not possibly have built such huge structures. Repeated attempts both by the church and by individuals to remove the giant stones failed.

In the face of this real evidence, it is impossible to believe that all these megaliths were built by sheer muscle power. We have no physical proof that there ever existed people of extraordinary size and strength who could have built the megaliths. The skeletons we have found of people who may have had a connection with the megaliths indicate that they were very similar to ourselves.

Sheer size is not the only peculiarity of the megaliths: their builders also displayed mathematical knowledge of an extremely high order—*as high as or higher than what we possess today*. The experts tell us that at this time (the Stone Age) there were no written languages in use. Then by what method did they arrive at their extraordinarily efficient mathematical formulas and the design layout at Stonehenge, for example? Did they carry out their computations and design the structure in their heads, and calculate everything from memory? If so, people of such a high degree of intellect would not have wasted their years and energies on dragging huge stones all over the countryside.

We are faced here with the same dilemma as the "Stone Age" artists who had perfect memories, and who by means of this faculty should have progressed to civilization by leaps and bounds.

We cannot be sure that they did not possess a written language, or the means for recording, and that these documents have not survived the passage of time. Indeed, there is no reason to suppose that the megalith builders wanted to preserve their documents. When they had arrived at their answers, and solved their equations, they may have felt little need to preserve them for posterity. In five thousand years' time, if our culture were to be destroyed, there would be little or no trace of our literary attainments unless specific efforts were made to preserve them.

Future ages may neither understand nor realize the significance of any discovery of a language of which both the oral and written forms had long since passed into oblivion. Who can read the Maya glyphs, or Cretan script, or understand those stone figures discovered on Easter Island? These strange findings may hold answers to riddles which have long intrigued the mind of man.

Of course, it is possible that the megalith builders did *not* use a written language. They, like Paleolithic man, may have possessed eidetic memory. In either case, we are left only with the results of their calculations: arrangements of huge mute stones.

These are the relics of people as wise as—or wiser than—ourselves, whether they used a written language or not, which brings us to another point. *Was writing, like the wheel, one of the world's great inventions, or was it another stopgap?* The advancement of civilization is not the result of any increase in natural ability or intelligence so much as it is of an increase in the amount of accumulated information.

We are, indirectly, back to square one. We need recording devices to replace eidetic memory, which ancient man did not need. We have no more information available to us than did ancient man, except that we store it artificially, while he stored it naturally. Perhaps the loss of this faculty following the catastrophe was one of the factors that made the climb back to civilization such a slow process.

There are three main types of megalithic structures in the Old World, as distinct from the megaliths of the New:

1. The stone circles and other alignments of megaliths, the best preserved of which is Stonehenge;

2. The tombs, passage graves, and dolmens, common in Britain and Western Europe;

3. The great pyramidal construction in Egypt.

All appear to have been built at a similar time. They were built sometime *between 4000 and 2000* B.C. (no one has been able to date any of them exactly); *they may all have been built within a short period;* and *such things were never built again.* Later structures were merely copies on a much smaller scale. Were they all interconnected? If so, how? And *why* were such massive projects undertaken?

If we are guided by traditional points of view—that Stonehenge was some sort of sun temple, that the passage graves were just that—graves—and the pyramids were merely monuments to some ruler's megalomania—then they make no sense at all.

No one has been able to say with any certainty that the Great Pyramid, for example, was built either as a tomb or as a temple. But such an enormous and immensely strong structure must have been built for some good reason. What?

The same thing applies to Stonehenge. According to many authorities, it was built, at least in part, for astronomical purposes. Gerald S. Hawkins, in his book *Stonehenge Decoded,* suggests that Stonehenge was an extremely accurate astronomical device for measuring earth-moon-sun positions, that it was, in fact, a giant computer. Even so, prevailing opinion holds that the *reasons* for its construction were mainly religious, and had to do with a priestly caste and their ability, through Stonehenge, to predict eclipses and thus hold a sense of power over the local population.

Such reasoning, however, leaves much to be desired. People of such high mathematical and technical ability would surely not go to such lengths to satisfy the whim of a priestly order. Moreover, there is no evidence of religious symbolism whatsoever, not even a representation of a deity—perhaps the Earth Mother cult, so much in evidence in late prehistoric and early historic times.

To all intents and purposes, Stonehenge appears to have been purely functional. As the function was astronomical, connected with the relationship between the earth, moon, and sun, there must have been a very serious reason for the builders to go to such lengths.

Authorities have suggested that the dolmens and so-called passage graves were constructed as tombs. Here again, why did they go to extraordinary lengths? Why did they build massive walls, capped with roof blocks many feet thick and weighing, in some instances, hundreds of tons, then cover the whole thing with a deep layer of soil, merely to bury the dead? No race of historic times ever went to such lengths to bury the dead. Might there have been a very different reason for the construction of these immensely strong tombs? Were they built not as tombs but as shelters?

Stonehenge, on Salisbury Plain in England, is the greatest single megalithic structure in Europe. Through the centuries it has been thought to be many things: a Druid temple, a marketplace, a court of justice, a meeting place of kings, a sun temple, an entrance to the world of the dead, an ancient center of mystic arts, a remnant of vanished Atlantis, even a rendezvous point for the crews of flying saucers. Most often, it has been thought to be some sort of astronomical observatory, although always in connection with sun worship, or designed to enhance priestly power. Even as an observatory, the emphasis has always been religious. What if it were not so?

The actual construction of Stonehenge, according to the archeologists, follows roughly this pattern:

At at unknown date, thought to be somewhere about 2000 B.C., a great, perfectly circular ditch was dug. The ditch material surrounds the ditch as a bank, 380 feet in diameter, 8 feet wide, and 2 or 3 feet high. The inner bank is a circle 320 feet in diameter, 20 feet wide, and 6 feet high, made of glaring white hard chalk.

The ditch/bank is open to the northeast, forming an entrance to the circle; inside the entrance there are four little holes. Just inside the entrance gap there are two large holes, which appear to have held upright stones.

The now famous "heel stone" stands 100 feet outside the circle, slightly southeast of the line of the entrance. Fifty-six "Aubrey holes" stand just within the inner bank.

The heel stone is about 20 feet long, 8 feet wide, and 7 feet thick, and weighs approximately thirty-five tons. Its lower end is buried 4 feet in the ground. It is made of sandstone, called sarsen, and probably came from Marlborough Downs, twenty miles from Stonehenge.

The fifty-six Aubrey holes are carefully spaced out over a diameter of 288 feet, with a 16-foot interval between their centers. The greatest error in interval is 21 inches; Gerald Hawkins has pointed out that spacing of such accuracy over the distance involved was no mean engineering feat. The holes are between 2½ and 6 feet in width, 2 to 4 feet in depth, and are filled with chalk.

This "first phase" of Stonehenge, then, was a ditch, with an inner and outer bank forming large circles, with fifty-six holes spaced around inside the ditch, three standing stones, and four smaller holes which may once have held either stones or wooden posts, also within the circular ditch. The complex is oriented, by alignments and an entranceway, toward the midsummer rising sun.

There are also what have been called the four station stones, which form a rectangle in the central part of the circle formed by the fifty-six holes. These station stones may have been erected during the "early" phase, and are numbered 91, 92, 93, and 94. Only two, 91 and 93, are still in position.

The rectangular placement of the station stones intersects the center of the circle, the short sides lined up with the center heel-stone axis and the long sides, almost exactly perpendicular to the axis.

The "second phase" was the erection of the megaliths. Eighty-two bluestones weighing five tons each stand in two small concentric circles around the center of the enclosure, six feet apart and thirty-five feet from the center. This circle seems to have been designed as a pattern of radiating spokes of two stones each, but the double circle was never completed. Holes are missing on the west side and two holes at the entrance were only partially dug and no stones placed in them. On the southeast side a large circular depression was found, which may have been intended for a large, flat-topped stone.

The avenue of Stonehenge is forty-seven feet wide, lined by two parallel banks, and runs along the sunrise line a third of a mile into the valley, then curves to approach the river Avon at West Amesbury.

In the "third" phase of Stonehenge, the double circle of bluestones was taken down, and their whereabouts are not presently known. They were replaced by more than eighty-one huge sarsen stones from Marlborough Downs, and were erected in the same general area as the bluestones, but in a different pattern.

Around the center of the monument stands a horseshoe of five trilithons (two upright megaliths capped by a crosspiece or lintel). A circle of thirty uprights, joined across the top by lintels, encloses the trilithon horseshoe. The horseshoe opens to the northeast, oriented, like all constructions at Stonehenge, to the midsummer sunrise.

The trilithons are of different heights, increasing in size toward the center of the horseshoe. They vary from 20 to 24 feet in height, including crosspiece. The central and largest trilithon, before it fell and broke, measured 25 feet; its companion, 29 feet 8 inches long, is buried deeper in the ground so that an equal height shows above the surface. This largest stone weighs approximately fifty tons.

The crosspieces were not simply dropped onto the tops of the uprights. On the top of each upright there is a knob or boss, and a corresponding depression in the underside of the lintel fits exactly over the boss. Cabinetmakers call this the "mortise and tenon" system.

There were two circular series of holes outside the monument proper, one outside the other; the inner group was called the Z holes, of which there are twenty-nine, and the outer, the Y holes, thirty in number.

Many aspects of Stonehenge suggest it was obviously more than a crude work of stone placing:

In the trilithon horseshoe, the uprights are carefully shaped to create the impression of straightness. This is achieved by tapering, some in a slightly convex curve, toward the top. This highly sophisticated architectural technique is called entasis. The lintels are similarly treated to create the visual impression of straightness by having their edges widened by some six inches. Their circumferential surfaces are curved slightly inward, the outer surfaces more curved than the inner.

The stones comprising the sarsen circle which enclose the horse-shoe weigh some 25 tons each, as opposed to the 45 to 50 tons of the horseshoe uprights. The lintels of the sarsen ring weigh 7 tons each. The uprights are 18 feet long, 7 feet wide, and 3 feet thick. Their bottom 4 feet are buried in the ground. As with the trilithons, both the uprights and the lintels are shaped to produce an illusion of straightness.

The sarsen circle is carefully spaced: the circumference is 97 feet 4 inches in diameter, and the thirty uprights were spaced with a maximum error of 4 inches apart. To the northeast, again on the mid-summer sunrise line, there is an entrance made by spacing two stones 12 inches farther apart than the others.

The center of the sarsen circle does not coincide exactly with the center of the older, original circle. If it did, the sun could not be seen to rise through arch 30-1 above the heel stone in midsummer. This displacement must have been deliberate.

One final alteration was made at Stonehenge: twenty of the bluestones, which had been removed when the sarsen circle was erected, were placed in an oval formation within the horseshoe. Later, these stones were removed.

The so-called altar stone (16 by 3½ by 1¾ feet) lies in the earth some 15 feet within the central sarsen trilithon complex, but its original position is not known and cannot even be estimated. Unlike all the other stones at the site, which are either bluestones or sarsen sandstone, the altar stone is made of fine-grained green sandstone, with many flakes of mica in its surface, so that freshly exposed surfaces glitter.

Just three quarters of a mile from the center of Stonehenge, Miss E. V. W. Field recently discovered a deep shaft—a 20-foot funnel-shaped depression that tapers into a hole 6 feet wide and at least 100 feet deep.

The sarsens at the site were brought from Marlborough Downs some 20 miles away. The bluestones were brought a much greater distance, from the only place in the British Isles they could possibly have come from, namely, the Prescelly Mountains in Wales, 130 miles away as the crow flies, and over 200 miles by any overland or sea/overland route. The green sandstone altar stone was brought from the Cosheshton Beds at Milford Haven on the coast of Wales, some 30 miles to the southwest of the Prescelly Mountains.

In the bottom of each of the Y and Z holes, it appears that a fragment of bluestone was placed; no explanation, or guess, has ever been advanced as to why this was done.

At the present time, a six-foot man looking from the center of the complex sees the top of the heel stone level with the horizon. In 1000 B.C. the first flash of the sun appeared three quarters of a degree to the north—provided the stone was inclined to the angle it has today. If, however, the stone was then in its true upright position, the error of 0.°5 would have been zero. (Opinion expressed by Gerald S. Hawkins.)

The degree of precision is extraordinary: this heavy and unevenly shaped boulder had to be placed in a hole in the ground exactly in the right position. If placed too high, it would have to be removed and the hole deepened. If placed too low, it would again have to be removed. Even if it had been dropped into exactly the right position, its weight (thirty-five tons) might well have made it settle lower. No one could be certain how much it would eventually settle.

Such precision at Stonehenge was not limited to the heel stone. Some of the fallen stones were re-erected in 1958, with the aid of extremely powerful cranes. Even with these modern machines, it proved impossible to align or set them up as accurately as the ancient builders had.

Stonehenge, it seems, was designed to record precise movements of the sun and the moon. Newham and Charrière have noted that the latitude of Stonehenge (51.17° north) is practically optimum for accurate rectangular sun-moon alignments. If it had been built on a site north or south by as little as thirty miles, the structure would have had the shape of a parallelogram, not a rectangle. The farther south it was placed, the more the figure would have changed.

It seems unlikely that the Stonehenge builders chose this particular site arbitrarily; they must have had a good reason. They may have chosen it because it was the only site—possibly in the world—where such a construction could have been undertaken. Nowhere else in this particular latitude, perhaps, was the terrain suitable. If they were aware of this, and of the mathematical fact that this latitude was the correct one for their purposes, were they not also aware of the size and extent of the planet as a whole?

Another point has been raised regarding the height of the uprights at Stonehenge. If they were needed only as sighting lines, small posts would have done equally well. Why such tall stones?

According to the calculations of astronomer Gerard Vaucouleurs, the positions and heights of the stones are such that at midwinter noon the shadow of the southernmost lintel of the sarsen circle falls precisely in the center of the complex. At midsummer, the shadow of the sarsen circle lintels falls on the bluestone circle. It appears likely, then, that these heights were fixed deliberately to create such shadow effects.

Dr. Gerhard Wiebe of the Boston School of Public Communications has commented: "Stonehenge makes no sense when seen from the ground. It is impressive only when seen in a plane from above. But Neolithic man had no airplanes from which to view his own work; therefore he may have been signaling his prowess to the powers in the sky . . . to his gods." He added that the great "Serpent Mound" near Peebles, Ohio, could only be appreciated from the air.

These two sites are not the only works of man that make no sense unless seen from a great height. The famous "lines" on a desert plateau at Nazca in Peru were not even discovered until aircraft flew over them. It has been theorized that the Piri Reis map of Antarctica was drawn from photographs or plans made at a great height; Von Däniken has suggested that the distortion shown on the eastern coast of the South American continent is consistent with observations taken from a height of eighty miles.

Why should the ancients have built things which could be seen properly only from the air? Could they have been executed with such precision to begin with—unless there was a way of checking from a great height?

Are we assuming too much when we baldly state, "Neolithic man had no airplanes"? Perhaps we should rid ourselves of the idea that some primitive "Neolithic man" actually built these constructions. Is Neolithic man a fantasy we have dreamed up ourselves, who never actually existed? Can we be sure that the megalith builders did not possess some sort of flying devices?

According to one of the legends of Indians in South America, the gods punished man in the great Flood because he had learned to fly. We remember the "vinamas" of Vedic literature, the ancient Egyptians whose gods flew in "boats" in the sky, and the *Arabian Nights* flying carpet. Are these echoes of a time when flying devices existed?

Let us return to Stonehenge:

On the basis of computer syntheses of a great many measurements at Stonehenge, and certain astronomical data, Gerald Hawkins has suggested, in *Stonehenge Decoded*, that *Stonehenge was, in fact, a computer for measuring all the relationships between earth, the sun, and the moon.*

Of his findings, several are especially relevant:

The fifty-six Aubrey holes have always been a puzzle. Hawkins says they could have been used as a computer, as follows:

> If one stone was moved around the circle one position, or Aubrey hole, each year, all the extremes of the seasonal moon, and eclipses of the sun and moon at the solstices and equinoxes, could have been foreseen. If six stones, spaced 9, 9, 10, 9, 9, 10 Aubrey holes apart, were used, each of them moved one hole counterclockwise each year, astonishing powers of prediction could be achieved.
>
> With six stones, three white, three black, the Aubrey hole computer could have predicted—quite accurately—every important moon event for hundreds of years.

The sun moves from a northernmost maximum position of $+23°.5$ declination in summer to a $-23°.5$ extreme southern declination in winter. The reverse happens with the moon; it goes north in winter, south in summer. The moon also has a more complicated relative motion than the sun, in that it has two northern and two southern maxima. In an 18.61-year cycle it swings so that its far north and south declinations move from $29°$ to $19°$ and back to $29°$. This motion is caused by the combined effects of tilt and precession of the orbit of the earth.

The Stonehenge complex was accurately aligned to all these movements. If we observe the 18.61-year cycle, we note that the only way to establish accuracy with whole numbers is by the method 19–19–18, which added together, make 56—*the number of Aubrey holes.*

The plain fact is that Stonehenge is a brilliantly designed and constructed astronomical device, which could calculate every position of the sun and moon.

The question is, WHY?

From the mathematical facts we have gleaned from Stonehenge, we can be certain the Stonehenge builders, since they were able accurately to predict the summer and winter solstices, must have known the exact length of the year, even to the quarter day.

From the angles of the sun, which could be checked using both the heel stone and the trilithons, and using the shadows cast by the pillars, it could be deduced that they were aware of the degree of the axial tilt of earth. If they were aware that the earth was a sphere, they may also have been aware that earth orbits the sun. In the case of the moon, the builders of Stonehenge would have known both the exact period of a lunation and the eccentricity of its orbit.

This is perhaps the most significant peculiarity of Stonehenge: the fact that it was constructed apparently to measure only the temporal and angularity sequences of the sun and the moon.

> In addition to complex computations of the sun and moon, the Maya made complicated charts of the synodal year of Venus. The ancient Egyptians devised astronomical charts by reference to many of the fixed stars, particularly the pole star. Babylonian and Assyrian astronomers were also interested in the constellations.

Most ancient astronomical observatories were also concerned with the stars. The megalithic structures of Western Europe, on the other hand, had to do only with the sun and the moon.

Why?

Although the general view is that Stonehenge was built around 2000 B.C., no one is really sure of this. The "first" phase of Stonehenge could be much older. We have suggested that all of Stonehenge was from its inception the result of a single project of short duration. A date of 4000 B.C. is not unreasonable and, in view of other evidence for the period known as the Flood, may be nearer to the correct time.

The Flood, one manifestation of a vast catastrophe, had, we suggest, a great deal to do with the construction of Stonehenge and other megalithic structures. In fact, we believe, Stonehenge was built *to determine the new orbital position and, accordingly, length of year, and the new axial position of earth following the catastrophe.*

If, as we have suggested, the earth had been abruptly jolted from its previously held orbital position, the distance it had actually moved, on an astronomical scale, would only have been slight. Even a slight movement away from the sun would, however, have been disastrous. Raising or lowering the average temperature of the planet only a few degrees would have the most appalling climatic consequences.

Such a jolt would also have severely affected the moon. If the moon happened to be in a position on the side away from the sun, for example, a change of the earth's orbit away from the sun would have moved the moon nearer to earth.

The original ancient calendars measured a shorter year (360 days), which suggests that earth was slightly nearer to the sun; the lunation, or lunar month, was 36 days, which means that the moon orbited the earth at a greater distance.

The jolting of earth would have reversed this situation: earth would have been farther from the sun, making a longer year, and the moon would have orbited nearer to earth, making a shorter lunar month. *Revised ancient calendars reflect this change.*

Is it coincidence that there are many legends of a great catastrophe that led to a re-evaluation of the length of the year and the lunar month, and the construction of Stonehenge and other megalithic monuments to measure precisely the movements of the sun and moon?

Further evidence supporting this point of view concerns the builders of the megalithic computing devices:

We have theorized that after the catastrophe most of the survivors fled, founding new settlements in areas away from the zones of greatest destruction, one of which was the northern region of the northern hemisphere. Surviving mathematicians and scientists would have realized that changes in the length of the year and the climate of the entire planet would follow. Accordingly, they made preparations to obtain data on the nature and extent of the catastrophe and the new conditions that would prevail.

To this end, it is possible that groups of such scientists either stayed behind to observe or returned to the devastated area shortly afterward, sheltering themselves as best they could from the effects of the catastrophe. The site they chose for the central computing installation—Stonehenge—was well within the most severely affected region.

Their megalithic computers covered the whole of Britain from northern Scotland and the Scottish islands to Salisbury Plain, and western France. It is possible that these complexes formed a huge set of astronomical devices with Stonehenge as the central unit, against which the findings of all the others were correlated.

After the final phase of Stonehenge was completed, it is generally agreed, the site was abandoned for many centuries—until the Druids began to use it as a temple. No doubt the same thing applied to the other megalithic constructions. This aspect has often puzzled archeologists and prehistorians. If, however, the megaliths were computing devices built to perform certain specific functions, it makes perfect sense. Once their purpose had been accomplished, there was no further need for them, and they were left.

Who were the mysterious builders of the megaliths? We have called them the surviving mathematicians of the cataclysm, the remnant of pre-Flood science. Perhaps they were the giants of antiquity, the "mighty men of old" of the Bible, the "culture bearers," the gods who brought wisdom and mathematics to the survivors.

Among the gifts the "gods" of legend brought to these rebuilt civilizations were the science of mathematics and the calendar—the new, altered calendar of 365¼ days. Greek legend may give us our clearest view of the originators of Stonehenge and hence of the other megalith builders.

We are taught that most of our basic scientific ideas, including mathematics, derive from Greece. But whence came Greek knowledge? Much of it—especially medicine, hydraulics, and agriculture—came from Egypt.

In the field of mathematics, however, early Greek writers give the credit to the elusive Hyperboreans, who, according to Greek historians, lived in a far northern land beyond where *boreas* (the north wind) blew. This land of the Hyperboreans is thought to have been Britain.

Diodorus, the Greek historian of the first century B.C., frequently mentioned the Hyperboreans and *their great round temple:*

"This island . . . is both fertile and productive of every crop, and since it has an unusually temperate climate it produces two harvests a year. Moreover, the following legend is told concerning it: Leto [mother of Apollo and Artemis; Zeus was their father] was born on this island, and for that reason Apollo is honored among them above all other gods, and the inhabitants are looked upon as priests of Apollo. And there is also on the island both a magnificent sacred precinct of Apollo and a notable temple which is adorned with many votive offerings and is spherical in shape.

"The Hyperboreans also have a language peculiar to them, and are most friendly disposed toward the Greeks and especially toward the Athenians and the Delians, who have inherited this good will from most ancient times."

The Greeks frequently mentioned the fact that the Hyperboreans taught them mathematics and that they studied with the Hyperboreans.

Representations of Mycaenean weapons have been found carved on some of the Stonehenge megaliths. According to some authorities, these carvings prove that Stonehenge was actually designed and built by master builders from the Mediterranean, since, it is claimed, the people of Britain were insufficiently advanced to have done it. In view of the Greek statement that Greeks came to the round temple to study mathematics, does it not seem more likely that these carvings were made by pilgrims to this place?

There are, furthermore, Greek legends about the Hyperboreans and flying. The Greek poet Pindar wrote, "Neither by ship nor by land canst thou find the wondrous road to the Hyperboreans." If not by sea and land, then how? Does this cryptic statement mean that the road was through the air?

Another Greek legend tells of a Hyperborean called Abaris, who rode on "Apollo's Arrow" through the air to Athens and Delos. Herodotus, when he mentioned this, did not believe it. Does this refer to some kind of flying device? A high-speed aircraft, seen from the ground, resembles an arrow.

A considerable body of legend stemming from the Mediterranean region concerns this northern land and its mysterious inhabitants, the Hyperboreans, and their round temple. The Hyperboreans were mathematicians and astronomers who were devoted to the study of the movements of the sun and the moon. There was only one round temple in "northern lands" that was devoted to astronomy, especially the movements of the sun and the moon: *Stonehenge*.

Most other megalithic sites in the British Isles and western France, probably closely connected with Stonehenge, have never been investigated. The first thorough measurements at Carnac in France were undertaken only in 1970, and it may be some time before the findings are analyzed and published.

In the vicinity of Stonehenge itself are several other sites, including Woodhenge, on a much smaller scale than Stonehenge, and apparently also used for astronomical purposes; and, seventeen miles north of Stonehenge, Avebury, a huge complex of stones, many of which have disappeared.

The circular shape of the Avebury site can now be discerned only from the air. Apparently it was composed of two rings of standing stones, 320 feet in diameter, their outer edges 50 feet apart, their centers oriented on a northeast/southwest line. A great ditch, with a diameter of 1,250 feet, contained a circle of 100 sarsen stones, the largest weighing over 40 tons. Avebury has until now not been thoroughly investigated.

Silbury Hill, sixteen miles north of Stonehenge, is the largest artificial hill in Europe. It has been called the "great pyramid" of Europe. The hill has a circular base more than 200 yards in diameter and rises to a height of 130 feet, covering some five and a half acres. More than a million cubic yards of chalk were used in its construction, and although both vertical and horizontal shafts have been dug through the mound, nothing of significance has been found. Whatever the mound may have been built to conceal—if anything—is either well hidden or lies perhaps in a chamber beneath the surface level of the mound.

A prehistoric road called the Ickneild Way, runs two hundred miles from Salisbury Plain to the Norfolk coast above London. In places it can still be seen to have been as wide as a four-lane highway. It is interesting to speculate whether this highway runs under what is now the North Sea, and whether any trace of a roadway which may connect with it can be found on the European continent. *Why should prehistoric peoples have built such broad roads?* These faint traces may be all that remain of a great highway system spanning the land mass that once included Britain, the continent of Europe, and what is now the bed of the North Sea, before the catastrophe.

There are many stone circles scattered throughout the British Isles, of which the most elaborate is located at Callenish on Lewis Island in the Outer Hebrides. Callenish is a ring of thirteen stones with a large central stone, an avenue, and other set rows of stones. Investigations have determined that there are ten alignments with the sun and the moon at their extreme positions on the horizon.

The latitude of Callenish lies near the Arctic Circle. Because of its location just south of the moon's extreme declination below the southern horizon, the full moon at midsummer stands 1° above the horizon once every eighteen or nineteen years. At midsummer, the moon sets over Mount Clisham, and the avenue points to the mountain. The nineteen-year cycle appears to have been known to the builders of Callenish.

The greatest megalithic complex apart from Stonehenge lies on the coast of Brittany in France, and is named after the small French town of Carnac.

Less than a mile northwest of Carnac is a huge semicircle of seventy closely spaced stones. Leading to the semicircle from the

> The Greek Meton is credited with the discovery of the nineteen-year cycle in 432 B.C., but the knowledge was not put to use until 312 B.C., during the Seleucid Empire. Meton may have stumbled across some of the knowledge left in Greece by the Hyperboreans.

southwest are eleven parallel rows of 1,100 columns. The rows are 100 yards wide and 1,100 yards long. These columns, called menhirs, increase from 2 to 12 feet in height as they near the semicircle. Three hundred and fifty yards to the east-northeast of these rows of columns are ten rows of a thousand stones, each row 1,300 yards long.

There are a further thirteen rows in a column 140 yards wide and 900 yards long, but of the original number of menhirs, which is not known, only 555 remain.

Although nothing is known about the site as yet, it seems that, like Stonehenge, Callenish, and others, Carnac's functions were mathematical and astronomical. The orientation of Carnac is the same as that of Stonehenge and Callenish—a southwest/northeast axis—and the astronomical connection is obvious.

Alexander Thom, emeritus professor of engineering at Oxford University, has investigated the ancient stone circles throughout the British Isles, 140 of which are in good enough condition to be studied.

Thom has found that those that were not perfect circles were not haphazard arrangements of stones, but geometrically exact designs. One side was an exact semicircle and the other was a flattened or bulging approximation of the circle. In some cases their circumferences equaled almost three times their diameters. For one, the ratio was 3.059, for another, 2.957. A true circle is 3.141596—a number which cannot be written exactly.

Were the ancients trying to make circles whose ratio equaled exactly 3? A modern engineer, working on the scale of these stone circles, Thom reports, would have been no more accurate. Indeed, the ratios are so minute it is possible that when the circles were new some of them did equal exactly 3.

Thom also found that many of the egg-shaped circles were so constructed that lines joining the centers from which the arcs were swung formed right triangles. Some of the circles are not modified circles; they are true ellipses, which are extremely difficult to construct.

Concluding his investigations, Thom asserts that the builders of these megaliths had a good working knowledge of geometry, including Pythagoras' theorem. The builders of the British megaliths (the Hyperboreans) may have given Pythagoras his theorem!

To summarize:

All these megalithic complexes are connected with astronomy, and some of them at least are connected with sun/moon alignments. Those that have been investigated are all oriented on a northeast/southwest axis. It seems that there is a connection among all these megalithic complexes, and that the same group, or groups, of people were responsible for them.

We suggest that *all these megalithic complexes were part of a vast mathematical and astronomical undertaking, built to chart the relationship between earth, the sun, and the moon. Their joint purpose was to calculate the degree of shift the earth had suffered in its orbit around the sun.*

Perhaps Stonehenge was the focal point for this great scheme. Perhaps the rearrangements that took place at Stonehenge were not undertaken by different groups at different times, but resulted from a series of mathematical measurements and experiments. As one scheme was completed and its data noted and evaluated, the circles were rearranged for the next series of calculations.

If all these constructions *were* part of one great complex, what a staggering picture it conjures up: a computing device covering many hundreds of square miles of country, spread across part of a continent and a group of offshore islands. This computer, besides being almost incredibly accurate, was virtually foolproof, unlike modern computers, which rely on electrical energy, electronic parts, and human programing. Here we have a huge computing device, which has no moving parts, which functions through the movements of the solar bodies themselves, which, once in place, cannot err, cannot wear out, cannot break down.

When one thinks of the energy, skill, planning, and intelligence to mount such a project, the efforts of our own science seem like child's play by comparison. The "giants in those days" to whom the Bible refers were not physical giants, but giants of intellect—perhaps almost the last representatives of a great, dying civilization that seemed to later generations an age of gods and magic.

How were these megaliths built?

How these widely separated groups of mathematicians and engineers communicated we do not know. Do we perhaps have a clue in the myth of the Hyperboreans and their use of "Apollo's Arrow"? A high-speed aircraft would have been an ideal method, not only of maintaining fast and efficient communication, but also of checking progress of the work from the air.

Did they perhaps use something akin to our radio equipment? Parts of ancient batteries have been found; if electrical energy was used, it is possible that radio devices were also known. Or perhaps these people with eidetic memory (which would have enabled them to store a great deal of the necessary information in their heads) were telepathic.

Looking at these battered stones today, it may seem odd to think of the builders using such modern devices as aircraft. But we have demonstrated the degree of intelligence possessed by the builders in designing these complexes. If they could do what is beyond our modern abilities, even with computers and heavy moving machinery, why could they not have used aircraft?

The collapse of a technologically superior civilization would have left little time for salvaging anything but essentials. The survivors may have salvaged certain needed devices, among them a few aircraft, which would have been required to maintain contact with other survivors. In the course of time, it would have become more and more difficult to keep such machines in working order. Parts would wear out, sources of fuel and power would fail.

If the technology is destroyed, the capacity for manufacturing machine parts, or even making the right metals, will no longer exist. (If, for example, a modern electronics engineer were transferred in time to ancient Rome, he would still have the knowledge to build, say, a tape recorder; but without materials, he could spend a lifetime trying and not succeed.) When such aircraft of the ancients ceased working, they would not long survive the passage of time and the assault of the elements.

If our civilization were to be destroyed, such things might well come to pass. In the end, all that would be left would be a memory of strange, dart-shaped flying things in which people once traveled through the sky. Ages later, who would believe such fantastic stories?

The story of the flying men is a part of a widely diffused legend. The ancients did not write these things simply to confuse our modern experts—though one would almost think they did, judging from the reactions of some experts—nor as fairy tales, but as things dimly remembered.

Many legends from the remote past are only now beginning to make sense. Who, in the early part of the twentieth century, would have associated a flying arrow with an aircraft? Air machines of this shape did not exist. The descriptions of nuclear devices in the *Mahabharata* made no sense to anyone until nuclear weapons were developed in the twentieth century.

Such connections were not made *because the devices did not exist*, and could therefore not be imagined. Can we say that the recorders of these prehistoric legends imagined the unimaginable things of which they wrote?

How such megaliths were elevated has been the subject of much speculation, none of it satisfactory. Some have suggested that the lintels at Stonehenge, for example, were raised on a rising latticework of logs. Such a latticed tower of logs has been estimated to have had to contain a mile of six-foot logs cut into twenty-foot sections. If we assume the builders had only stone tools, this alone would have been a tremendously time-consuming undertaking.

Others suggest that an earthen ramp was built to drag the lintel up; but the task of building up and then removing a ramp for each of the thirty-five lintels at Stonehenge would have been a more stupendous task than the digging of the original circular earthwork. No evidence indicates that such earth-moving operations were ever undertaken. Wooden ramps would have involved as much timber as a latticed tower and been much more dangerous.

All these methods seem hopelessly impractical in connection with megalithic building. Apart from the seven-ton Stonehenge lintels, other megalithic structures in Britain alone present even more baffling construction problems.

For example, there are the *dolmens*, consisting of three uprights placed in a triangle, with a single slab resting on their points. Some of these roof slabs weigh hundreds of tons. Even with present-day mechanical equipment, we are unable to raise such weights.

The bluestones used at Stonehenge were brought from the Prescelly Mountains in Wales, a distance, as the crow flies, of 130 miles. The shortest overland and water route is 240 miles. More than eighty five-ton bluestones were used in the construction of Stonehenge. Could teams of men have dragged and hauled these stones on rollers and by raft all that distance?

In an experiment undertaken by the BBC to find out how the stones could have been moved by the manual method, it was found that sixteen men per ton could move a mile a day. Eighty men would have taken 240 days to move *one* bluestone from Wales to the site. If all eighty bluestones were moved at the same time, 6,400 men would have to haul for three quarters of a year. Over the rough ground and steep slopes that characterize the terrain between Wales and Salisbury Plain, there would have been days when virtually no progress could be made.

It could have taken a year or more, and the efforts of thousands of men, to move the bluestones alone to the site. *Why were these particular stones brought all that distance?* Since Stonehenge was built primarily for mathematical reasons, and the builders were logical men, there must have been a good reason to bring bluestones from Wales. Did these stones have a special property?

The *Mahabharata* tells us that ancient man may have experienced the effects of nuclear weapons; atomic energy may have been one of their sources of power.

Was the incidence of radiation in this particular part of Wales higher than normal because of the weapons used? Were the bluestones, in fact, slightly radioactive?

Bluestone chips lie in the bottoms of the Aubrey holes, which are the outer perimeter of the monument. Tree roots, which have torn apart many ancient structures in all parts of the world, have never done so at Stonehenge. Vegetation has never grown in the area of the megalith. Was there something in the bluestones that ensured that nothing grew there that could damage it?

In woods in Virginia, groups of oddly shaped stones have been found upon and around which *nothing will grow*. Did these stones —like the stones at the Stonehenge site—possess some property which caused this?

Gerald Hawkins has estimated that the total time taken to build Stonehenge alone amounted to some 1.5 million man-days—not including making rollers and boats to transport the bluestones, administrative work and supervision, and feeding, clothing, and housing the workers involved.

If this edifice alone had been built by the "heave-ho" method, many thousands of people must have been involved. In addition to construction workers, there would be planners and supervisors, those required to grow food and look after stock, wives, children, old folks, and those who were sick or otherwise not able to work.

The total population may well have amounted, all told, to a figure not much less than 75,000 to 100,000 people, living in the vicinity of the site for many years.

Yet we are told that in Neolithic times *the population of the entire British Isles numbered in the thousands, if not the hundreds.* The Great Pyramid poses the same riddle: armies of laborers from a nonexistent population, great constructions without a sign of where these people lived.

The megaliths were in fact built by only a few people, mathematicians, astronomers, engineers, building in a land which had only just emerged from the effects of the catastrophe.

19

PREHISTORIC FALLOUT SHELTERS

"Passage graves," as their name implies, are underground or partially underground passages or tunnels with side chambers leading off the passage. Their walls and roofs consist of massive slabs of rough-hewn rock; generally the construction is covered with a thick layer of smaller stones and earth, forming a mound.

Geographically, the passage graves are found principally in the northern latitudes—Britain, western France, Ireland, Scandinavia—although there are some in Spain, Malta, and the Balearics. We find none in the Middle East, Italy, Greece, India, or any other Asian country.

The concentration of passage graves is greatest in those areas that suffered the greatest impact of the catastrophe. The passage graves, we suggest, were not tombs, religious shrines, or places of burial. They were shelters.

Shelters of several types have been discovered: rock-cut, corbeled dry vault, megalithic chamber, and cyclopean "giant's graves." Another type, called a *dolmen* in France and a *cromlech* in Wales, consists of several upright walling stones with a single massive cap-stone. Today many of these shelters appear to be bare stone masses; originally, they were covered by mounds of earth.

The shelters appear to have been built at approximately the same time as the stone circles, although some shelters may be somewhat older. The shelter at Ile-Carn in western Brittany, for example, has been radiocarbon-dated to 4000 B.C., a date which fits well with our tentative date for the catastrophe.

In *The Riddle of Prehistoric Britain,* Beaumont, suggesting that the "passage graves" were actually shelters, refers to a passage in Isaiah 20:1, which relates to other statements we have already cited:

"Come, my people, enter thou into thy chambers and shut thy doors about thee; hide thyself as it were for a little moment, until the indignation be overpast. For, behold, the Lord cometh out of his place to punish the inhabitants of the earth for their iniquity; the earth shall also disclose her blood and shall no more cover her slain."

Does this passage refer to the so-called passage graves, the shelters?

They exist in many parts of Europe, *but not in the Middle East.* Isaiah is describing events that occurred in regions far removed from the Middle East. As we have observed in connection with other references to the catastrophe, the writer probably had access to either oral traditions or the traces of documentary evidence of the time of the catastrophe, records brought by survivors when they fled the northern regions and settled in the Middle East.

Skeletons found in "passage graves" in Scandinavia, on some of the Danish islands, in Britain, and elsewhere, appear to have been pushed to one side to make room for more corpses. Many "passage graves" also contain cooking vessels, tools, and implements. Occasionally there is evidence that fires had been lit inside them.

Traditionally, it has been maintained that these items were placed in these "graves" for the use of the dead in the afterlife, and that the fires were lit for some religious purpose. Is it not more possible that these things were taken into the "passage graves" by living people?

Generally speaking, the "passage graves" were built with a very small entrance, in some cases only three feet high, so that they had to be entered in a crouching position. Some of the slabs that covered these entrances had portholes cut into them. The passage inside, often stretching for over a hundred feet, gradually widened and deepened. Small chambers off the passage were large enough for a man to sit and lie in, but not to stand upright in. Within the "passage graves" skeletons have been found, widely scattered and in many different postures, together with artifacts and the remains of fires.

The small entrance, the massive walls and the even more massive roofs, and the thick outer covering all suggest nothing so much as an air raid shelter.

In Jersey, the dolmen of Grantez St. Ouen stands 212 feet above the sea. When it was opened, it was found to contain the remains of eight people: five lay on their sides in a crouching position at floor level; one, five feet distant, was in the entrance to the main chamber, below floor level; another was nine feet south of the last in the passage in a sitting position. In a side chamber were the remains of a child and a few scattered bones. In the main chamber were a stand for an urn, a shallow oval dish, and a saucer-shaped plate.

Does this sound like a burial chamber? Were people buried so haphazardly in such an elaborately created tomb? It sounds more like a place where living people were taking refuge from some great danger, were overcome, and died. Exactly what killed them we are unable to say.

Perhaps they were killed by a blast, perhaps they were drowned—in the case of the disaster on the scale we have envisaged, waves of this height would not have been unusual—or perhaps they died of starvation. Possibly they died from radiation sickness.

The dolmen at Grantez St. Ouen is typical of many cases in which it appears that at least some of the people in the shelters were alive when they died from the effects of the catastrophe. Afterward, the shelters may well have been used as tombs to bury those who had used them in life.

It seems reasonable to assume that later generations who returned to these areas buried their dead in these already existing structures, either because they were there or because some religious significance had become attached to them. Curiously, after about 1500 B.C., the shelters were no longer used for any purpose. The majority of the people inhabiting these areas were cremated; only important people—rulers, tribal chiefs, and important priests—appear to have been preserved.

What were originally shelters became tombs, in the same way

that, at Stonehenge, what was originally a computing device became a temple.

If our civilization were destroyed, future ages may make the same mistake and judge as sacred areas the remains of structures totally remote from any present religious associations: a computer installation, a railway station, a bank building, or the Statue of Liberty!

Apart from the "passage graves" and dolmens of Western Europe, similar structures—notably the temples of Malta—exist in more southerly latitudes. No satisfactory explanation has ever been offered as to how or why these deep underground megalithic chambers were built in solid rock.

The dolmens and "passage graves," involving as they did masses of stone far larger than those used in the construction of Stonehenge and the other circles, must have been built by the same methods. *In fact, it seems likely that all megalithic structures in every part of the world which date from this ancient period were built by the same advanced methods.*

20

The largest surviving stone structure of the ancient world, the Pyramid of Cheops, is one of some sixty mysterious stone structures stretching in a narrow band along the Nile Delta of Egypt. (Thirty pyramids have now practically disappeared except for outlines of foundations.)

Were the pyramids built as tombs for the pharaohs? Were they built by—or for—extraterrestrial visitors? In either case, how were they built?

The riddles surrounding the construction of the Great Pyramid are typical:

When was it built? The Greek historian Herodotus was told that it was built as the tomb for the Pharaoh Cheops, that 100,000 men were involved in its construction, and that it took twenty years to complete. This would place the building during the early part of the Old Kingdom, between 2600 and 2500 B.C.

Unfortunately, the only chronologies we have of the kings of Egypt are those recorded by Herodotus, based on what he was told by Egyptian priests, and a fragmentary history written by the Egyptian priest-historian Manetho. Neither mentions a king called Cheops.

In fact the Egyptians of Herodotus' time were as puzzled by the pyramids as we are today. The origin and purpose of the pyramids was then, as it is now, shrouded in mystery; no documentary evidence exists that accounts for how, when, why, or by whom they were built.

Like the "passage graves," dolmens, and stone circles, the pyramids were all built on a megalithic scale—and all appear to have been built where there is little or no trace of human habitation. Furthermore, it appears that the pyramids were built during the same time as the other megaliths.

The book *Prehistoric Europe* by P. van Doren Stern suggests that the entire population of ancient Egypt probably was not much more than 1 million.

When one considers the enormous size and number of the pyramids, it is remarkable that there is no evidence of the kind of urban development that could have supported the large work force needed to build the pyramids by manual methods.

The pyramids are thought by most people to have been built as tombs. In fact, very little trace of burial has been found in them. The only physical proof we have of the elaborate funerary customs of pharaonic burial comes not from the pyramids but from the Valley of the Kings, where tombs, including the fabulously rich tomb of Tutankhamen, were hollowed from the cliffs in the valley.

The pyramids may have been used as burial places by later pharaohs, in some instances a thousand years or more after they were built. It was common practice for a pharaoh to commandeer the tomb of a former ruler, particularly one whose reputation had suffered since his death, eliminate all his inscriptions, and substitute his own. In some cases, traces of deliberate erasure and substitution have been discovered.

The builders of the pyramids, we suggest, were the same people who built the European megaliths. The ancient Egyptians who later claimed the pyramids as their own were their descendants or survivors of the catastrophe who settled in this region south of the Mediterranean.

The so-called Great Pyramid of Cheops—and the others, on a somewhat smaller scale—manifest technique, skill, and mathematical ability on a scale equal to that displayed at Stonehenge, but far more refined and sophisticated, possibly because the pyramids were built in a different location for a different purpose. Nonetheless, the construction of the pyramids was connected with the catastrophic events which prompted the buildings of the shelters and the astronomical circles.

Masudi, an Arab historian, stated that the Great Pyramid was built by a king called Surid, who lived before the Flood, and who had been warned of the coming deluge. Surid, Masudi tells us, caused the pyramid to be built both as a refuge for the living and as a safe for the knowledge they possessed until the catastrophe passed.

Coptic legend states that Surid ordered the inscription to be engraved in words, together with "other matters," on the pyramid casing. Many ancient Arab writers have mentioned that at one time the outer, polished limestone casing of the pyramid was covered with inscriptions. Of course, this outer casing has now completely disappeared.

Those who maintain that the Great Pyramid was a tomb point to a great stone coffer in the King's Chamber, which is slightly wider than the Grand Gallery and must therefore have been built into the pyramid before the chamber roof was in position. The coffer, however, has no lid, and no trace of an interior coffin has been found.

This coffer may not have been a sarcophagus at all. The tomb robbers who were alleged to have removed the coffin, corpse, and valuables would hardly have taken a sarcophagus cover weighing several tons. Where then is the cover—or was there none, since it was not a tomb to begin with?

There is no reason to suppose that the pyramids were built to house the dead or were connected with any form of religion.

No reliable record indicates how long the Great Pyramid was under construction. Some estimate, using Herodotus' account, that it took twenty years. Other traditions state that it took Surid sixty-one years to complete the task, or that it took one hundred days. How long this or any other pyramid took to build, we shall never know.

Furthermore, unlike many later Egyptian structures, we have absolutely no knowledge of the methods or the tools used in its construction. Later rulers of Egypt were fond of erecting stelae or commemorative plaques, detailing the wealth lavished on their tombs and buildings, the quantity of stone, timber, furnishings, number of men employed, even the names of architects and engineers. There is not one hint regarding the pyramids.

In one instance we do have a name, in connection with the step pyramid at Saqqara. The architect was reputed to be the master builder Imhotep. Imhotep was also a mathematician, scribe, and physician, and was deified after his death. Descriptions of Imhotep remind us of the culture bearers—the intellectuals who survived the catastrophe and helped rebuild civilization.

Considering that the pyramids have been regarded as the tombs of pharaohs, it is remarkable that *in not one instance has the embalmed corpse of one of these rulers been found in any pyramid.* The pyramids make very little sense as unused tombs.

Within a maze of underground galleries beneath the step pyramid at Saqqara, archeologists Firth and Quibell discovered twenty thousand alabaster jars, which, it is stated, may have contained food offerings—for the dead, presumably. The so-called burial chamber at the foot of a ninety-foot-deep shaft was empty. The only sign of human remains was one leg bone, found by Howard-Vyse over a hundred years ago. Were twenty thousand jars for food buried inside a pyramid for the use of *one* dead man? It seems highly unlikely that any people capable of constructing this pyramid would have so squandered their resources.

Does it not seem far more probable that these jars of food were to be used by the living who sought refuge in these galleries and chambers, deep underground, securely protected by the virtually impregnable mass of masonry above their heads? This would explain why the jars did not contain the petrified remains of food—if they had contained offerings for a dead man, traces would remain, preserved by the desiccating conditions prevailing in Egypt. Ears of corn and flowers have been found in actual burial tombs, so we know such traces should have survived. It also explains why *no corpses were ever found in the pyramids.* If the pyramids were used as shelters, on the other hand, the living simply walked out when the crisis passed.

Furthermore, in the Great Pyramid, *air shafts were built into each of the chambers.* If it was built as a tomb to preserve the pharaoh for all eternity, the design would surely not have provided for an air supply. Far from helping to preserve the corpse, air would only have contributed to a more rapid decomposition. But an air supply *would* be essential for the living.

> The Great Pyramid is built on solid bedrock, and is hence virtually invulnerable to the severest earthquake.

The shape of the pyramids also suggests a function other than a tomb: Beaumont has pointed out that the pyramidal shape is ideally suited to withstand almost any kind of shock—a blast of any intensity from any direction, and missiles of any kind, natural or artificial. Huge rocks from a volcanic eruption, large meteorites, bombs, or high-explosive missiles, would, no matter where they struck, either be carried to the base by the sloping sides, or explode without causing major structural damage, because there is no roof area to sustain the major shock of impact. The angled sides would also enable it to withstand blasts or high-velocity winds, unlike a straight-sided edifice.

The main chambers, well shielded by an enormous weight and thickness of masonry, or situated deep underground, increase the safety factor enormously. In the case of the Great Pyramid of Cheops, the so-called King's Chamber above ground level, in the heart of the pyramid, is buttressed by a roof composed of fifty-eight beams weighing fifty-six tons each.

The Great Pyramid of Cheops measures some 800 feet on a side, and is more than 850 feet high. It is so accurately aligned to the four cardinal points that a modern compass can be checked against it. With regard to the alignment there is one further peculiarity: we have already observed that the striations and vitrifications on rocks associated with the Ice Age theory occur mainly on northeast-facing slopes. The pyramid complex at Giza is oriented on a northeast axis, and a line through the entire pyramid complex in Egypt shows this northeast alignment. These factors must be more than coincidence. Again, this demonstrates the connection between the pyramids and other megaliths like Stonehenge—and the connection with a catastrophe.

The geographical location of the pyramids may also be connected with a worldwide catastrophe. We have suggested that one of the principal areas affected was the present Arctic region. Egypt is at a distance great enough from the epicenter to be less affected; it is also almost the only area of the Mediterranean region which is fairly stable geologically—less subject, for example, to earthquakes. That the pyramids were designed to withstand great stress is borne out by the fact

450' HIGH see pg. 203

that they show no sign of having been affected by the great eruption of Thera in 1500 B.C.—when they were already ancient—which caused other great destruction in Egypt from tidal waves and earthquakes.

It may be instructive to examine in some detail the internal arrangements of the Great Pyramid of Cheops, most of which apply, in lesser degree, to the other pyramids.

The entrance is 48 feet above ground level on the north side. From the entrance a passage leads downward at an angle of 26° 41′ for 320 feet. At 293 feet it terminates in a level passage which leads to a subterranean chamber, called the Pit, hollowed out of the solid rock 101.5 feet below the foundations.

An ascending corridor opens from the descending passage sixty feet below the entrance. At this point a triangular plug was built into the roof which could be lowered to seal the ascending passage.

A horizontal left passageway 123 feet higher up the ascending passage leads to the Queen's Chamber, in which the pointed roof of massive blocks of stone is 20 feet 4 inches high. The ascending passage opens up into the Grand Gallery, 155 feet long and 28 feet high, constructed of enormous blocks of stone, some of which weigh fifty tons. From the lower end of the Grand Gallery a shaft extends to the lowest subterranean passage, known as the Well.

Access to the King's Chamber is a small passage 22 feet long and only 3 feet 8 inches high; it can be entered only by crawling. It seems unlikely, if this were a burial chamber, that access to the most important chamber should be so difficult and undignified. Furthermore, it would be virtually impossible to carry any coffin through such a confined space, let alone the heavy and elaborate casket of a king.

The airshafts for the King's and Queen's Chambers had their outlets at heights of 340 feet and 220 feet, respectively, above surface level.

There are five hollow chambers above the King's Chamber, four with massive flat roofs, and the uppermost crowned with tiers of fifty-eight-ton sloping slabs leaning against each other (relieving beams).

Does it not now seem far more likely that the Great Pyramid was designed not as a tomb secure against theft but as a shelter against any combination of forces it may have had to withstand? If it was merely for protection against robbers, why were so many dummy chambers of immense strength built in above the King's Chamber? If the pyramid could be sealed by the great granite plugs inserted at strategic points, no robber could possibly have gained access to the chambers. The chambers, buffered as they were, had been designed to withstand enormous shocks.

Additional protection was afforded by the outer casing. Originally, the outer casing of the pyramid consisted of a smooth limestone façade, polished to perfection and so finely masoned that a piece of paper could not be inserted between the joints. This finely worked outer casing was not mere ostentation: the polished slopes would ensure that objects striking the pyramid would slide easily to the base.

In the event of a flood or tidal wave, the water would have to rise forty-eight feet to reach the entrance, thereafter pouring down the descending passage to fill the Pit, before rising to meet the junction of the ascending passage, where its progress would be halted by the plug. The plugs, built with the same incredible accuracy as the rest of the pyramid, would have afforded an air- and watertight seal.

The only other way water could have entered the chambers would have been through the airshafts, over 200 feet above ground level. It is extremely doubtful that the water would have reached such a height, even during a severe tsunami (tidal wave), since by the time such a wave reached the vicinity of the pyramids from the coast, much of its force would have been spent.

An analysis of the construction seems to indicate that it was designed to be a protection against missiles, the effects of blasts or high-velocity winds, and floodwaters caused by earthquakes and tidal waves.

Modern engineers and scientists are agreed that we could not duplicate this construction feat today—even with the aid of our strongest machines and all our computerized mathematical knowledge. Yet the traditional view has it that the builders, with nothing but muscle, rope, ramps, and rollers, achieved what we are unable to do.

It has been estimated that the Great Pyramid is composed of some 2.6 million blocks, each weighing between two and a half and fifteen tons. In addition, there are the massive slabs lining the Grand Gallery; the passage and chamber were hollowed out of the solid bedrock on which the pyramid rested.

Further constructions were involved: a great causeway of stone from the Nile to the pyramid plateau, with large buildings at either end; a moat surrounding the pyramid; recently discovered boat pits lined from fine ashlar masonry as protection for the boats they contained. It is possible that the entire complex was surrounded by a wall. The wall surrounding the Step Pyramid, and traces of walls around the other pyramids, have never been satisfactorily explained. Perhaps they were built to offer additional protection against earthquakes.

Herodotus, when he surveyed the pyramids at Giza 2,400 years ago, observed that the building of the causeways must have absorbed as much labor, energy, and material as the building of the pyramids themselves.

Estimates of the labor force involved vary between 2,500 and 100,000 (the lower figure is a recent suggestion). If we assume that the pyramids actually took five years to build (the most popular estimate at the present time), and that the ancient Egyptians worked only during the hours of daylight—say, ten hours—then the task of quarrying, cutting, shaping, and smoothing the blocks by the methods traditionally envisaged would mean that *each day*, over 1,400 blocks would have had to be laboriously split from the quarry, shaped by hand, polished, and smoothed to a hairline accuracy, dragged, placed on rafts, ferried across the Nile, hauled up innumerable ramps, and put into place on the pyramid.

When all the subsidiary construction—the moat, causeway, pits for the boats, the ramps for each course, the passageways and chambers, and the outer casing—is considered, the mind boggles at the task these people had set themselves by muscle power. As in the case of Stonehenge and all other megalithic stonework, the pyramids were built without the use of the wheel, pulley, or block and tackle. Even before the actual construction of the pyramid, a chamber was *drilled through solid bedrock* at the bottom of a hundred-foot-long subterranean passageway.

It is simply not possible!

The traditional view of pyramid building is that ramps were built leading to each stage of the construction, and that teams of men hauled the blocks into position up these ramps and then placed the stones into positions accurate to the thousandth of an inch. Models of this system usually demonstrate this process in the lower tiers, near ground level, where it looks absurdly simple.

No one has suggested how this was accomplished as the workers neared the top, or what happened when they reached the top, which, in the case of the Great Pyramid, is 450 feet high. The quantity of ramp material and the amount of labor needed to reach the topmost layers are so vast as to stagger the imagination. Furthermore, each succeeding course of a pyramidal construction is much smaller than the one below. When the apex of the pyramid is neared, not only is the height from the ground enormous, with consequently hundreds, if not thousands, of sections of ramp spiraling round the edifice; there would scarcely be room for more than several men to work.

Let us assume that a block weighing seven tons (the blocks at the top were no lighter or smaller than those nearer ground level) was being hauled from ground level into its position near the top. If we refer to the Stonehenge experiment, we see that sixteen men are required to move each ton of weight a mile a day over reasonably level ground. In the case of pyramid construction, moving the weight up an inclined plane, can we assume that twenty men to a ton is reasonable? If so, to move a seven-ton block would absorb the energy of at least 140 men; the task would have to be tackled by teams in relays, as one team could not cover the height and distance in one effort.

We know the pyramid was 800 feet to a side at the base, and although it gets smaller the higher we go, the total distance to reach a point near the top, by means of spiraling ramps, will increase the total distance covered by many miles. As the top is approached, however, the area shrinks from 800 feet a side to 20, then 10 feet. The number of men required to place the block in position would not have been able to work in this restricted space.

Considering the vast amount of dressed stone and the refinement involved in the construction of the Great Pyramid, it would seem that it alone would have taken vastly longer to build than either Stonehenge or Avebury (estimated to have required—by traditional building methods—1.5 million and 2.5 million man-days of labor, respectively). Von Däniken, in *Chariots of the Gods?*, has suggested that if the workers had fitted the blocks at the rate of ten a day (which he considered an extraordinarily good rate), the task would have occupied 654 years. A task which had occupied so much time would surely not have been so easily forgotten by Egyptian historians. But, as we have said, *they had no knowledge of it or its builders.*

It is simply beyond belief that people would labor so long on a single project, and to suggest that years or even centuries of work were undertaken merely to bury a king is the height of absurdity. Let us suppose that construction on a pyramid commenced as the pharaoh took the throne; it could hardly commence sooner, because he would lack the authority to authorize the work. Perhaps before he had reigned more than a year or so, he was deposed. What if he died, or met with an accident, or was killed in battle? Very few pharaohs managed to survive for a great many years. Would his subjects preserve him until the sepulcher was completed? If so, how?

Other theories of how the pyramids were built are equally unsatisfactory. Some say that the stones were sawed into shape, asserting that saw marks are still visible on the blocks. It has always been assumed, however, that during the period of history when the pyramids were supposedly built, the only metals known were copper, or the alloy of copper and tin, bronze. According to this theory, then, the saws were made of bronze, fitted with a hard substance (perhaps ground granite). The saw teeth would have had to be punched with a metal harder than the saw blade itself—but they were already using the hardest metal known to them. Perhaps they did know of other metals, but no trace of them has ever been found. Nor have any stoneworking tools been found at or near these sites.

The pyramid builders may have used something akin to a laser beam to cut stone. If such devices had been used by the pyramid builders, vast numbers of blocks and beams could have been cut swiftly and accurately.

In the case of Stonehenge and other European megaliths, we have suggested that antigravity devices may have been used to erect them. As in the case of Stonehenge, the work may have been completed in a matter of months or weeks. *Speed was imperative.* While Stonehenge was built *after* the catastrophe, to measure its effects, the pyramids were built as shelters from the catastrophe.

One last curiosity regarding the Great Pyramid applies also to Stonehenge: no trace of any religious motif has been found on or in the pyramid. No statuary or inscriptions adorn its walls or chambers. Like Stonehenge, *the Great Pyramid appears to have been purely functional.*

The pyramids of Egypt stem from a period of "Egyptian" history about which nothing is known, and are therefore "prehistoric," as are Stonehenge and the other megaliths. We believe that it is merely coincidental that the pyramids and the ruins of the ancient Egyptian civilization occupy the same area. They do not belong together. *The pyramids have no definite connection with the ancient Egyptian civili-*

zation; they are not, in fact, typically Egyptian. They were built at one particular period of time, for one specific purpose, and when this purpose was completed, they were left to time and the elements, a mystery to baffle all who came after.

They became the focal point for a new civilization, which used them as a guide for their own stonebuilding and developed a complex mythology and religion around them, creating in the end a static art and architecture that has lasted thousands of years.

The ruins of ancient civilizations lie scattered throughout Central and South America from Mexico to Chile. The real names of some of these cultures are unknown to us; others we know as Olmec, Maya, Zapotec, Toltec, Mixtec, Aztec. But there are nameless ones: the builders of the great city of Teotihuacán in Mexico, the complex at Cocle in Panama, the ancestors of the Inca, Chimu, Paracas, Nazca. Hundreds of ruined cities in the Andean jungles have never been explored.

As in the Old World, all these civilizations have abrupt beginnings; there are no apparent transitional stages. The older the civilization, the more advanced the techniques of construction seem to be.

More than the ancient civilizations of the Old World, those of the New have provoked widespread speculation. Ancient Egyptians, Israelites, early Christians, even Norsemen have all been brought into the puzzle.

Archeologists were bound to seek connections between Old and New World civilizations; indeed, many remarkable similarities do exist: pyramids, similar techniques of masonry construction, the Flood myth. The similarity between Inca and Roman methods has led scholars to call them the Romans of the Americas.

South America, like the Old World, has its quota of archeological riddles:

• The so-called fortress of Sacsahuaman, overlooking Cuzco, the ancient Inca capital, one of the most remarkable pieces of stonework in the world.

• The city of Tiahuanaco, on the shores of Lake Titicaca in Bolivia, which has raised considerable controversy as to its age and building methods.

• The lines at Nazca, of unknown antiquity and purpose.

• The puzzle of the Paracas textiles, which cannot be duplicated on our best machines.

• The medical skills of the Chimu, which rival anything of the ancient Egyptians or Greeks.

But for all the similarities with the Old World, there are also great differences. The Inca, the most cohesive and advanced empire in the Americas, apparently had no written language, and the Aztecs used a pictorial language. The Maya developed the most complex glyph writing on earth, one which resembles no other.

This absence of written languages, or these written languages which cannot be translated, is one of the great stumbling blocks to understanding these divergent cultures. Their origins, diffusion, way of life, and thought are almost unknown, for we have their ruins and very little else. These ancient civilizations, as a consequence, remain strange, alien, and remote, without the sense of history and continuity we have with ancient Greece and Rome, and to a certain extent with Crete, Egypt, and Sumer.

We have no *Odyssey*, no Jason of the fabled voyage, although many of these ancient Americans may have made voyages as great, as interesting, and as daring. We have no great stories of conquest as we have with the Romans, although the adventures of the Inca expansion may be as great as anything in European history. We have no equivalents of Rameses, Thutmosis III, or Ahkenaten, but perhaps the great religious centers of Teotihuacán or Pachacamac produced men as great. Archimedes and Euclid and Pythagoras may have had their equivalent among the mathematicians and engineers of the once great

Maya centers, but these men, and all those up to the time of the Spanish conquest, are nameless, faceless, and unknown.

The civilizations of the Americas were isolated for several thousand years, up to the time of the Spanish conquest. Even among themselves they were isolated: the three great cultural centers of the Americas—the valley of Mexico, home of the Toltec and Aztec; Central America, home of the Maya; and the Peru-Bolivia coastal region, where a great variety of cultures flourished, culminating in the great Inca empire—were completely unaware of the existence of the others. Because they had certain things in common, particularly the legend of the white culture gods, however, they may have been originated by groups of the same people.

This historical and geographical isolation has meant that the legends about the origins of civilization have not been diluted by an overlay of tradition. Perhaps this is why it has been thought that if these culture bearers existed at all they are much more recent than those of the Old World.

This comparative "newness" ascribed to the American cultures shows in archeologists' dating of the American cultures. The Maya civilization, for example, is frequently estimated to have been founded around 700 or 800 B.C.; wooden lintels found in a temple at Tikal in Guatemala, however, have been radiocarbon-dated at 1500 B.C. These dates (which may not refer to the most ancient structures of the Maya) correspond to dates given for many Old World cultures.

Dating is even more uncertain for many of the cultures in Peru and Bolivia, some of which are thought to be as old as the Maya. A cache of sandals almost identical in pattern to those worn by the Peruvian peoples at the time of the Incas has been found in a cave in North America and have been tentatively dated as far back as 8000 B.C.

The peculiarities of many physical remains extant in the Americas are reminiscent of those in the Old World. Let us examine some of them in some detail.

Tiahuanaco lies on the shores of Lake Titicaca just across the Peruvian border in Bolivia. As nearly as we can ascertain, Tiahuanaco means "the Place of Those Who Were" or "the Place of the Dead." *It had been ruined and deserted for centuries before the first Inca attained ascendancy.* The Inca considered it a mysterious place, built by gods or magic.

According to most traditional pre-Columbian scholars, the Tiahuanaco culture arose about A.D. 1000 and lasted several hundred years before passing into obscurity. Curiously, however, the Inca empire was at its height in the early sixteenth century when the Spaniards arrived. The Inca Yupanqui launched his campaign to conquer Chimu several centuries before the arrival of the Spaniards. It is difficult to see how the Inca could have built an empire covering an area almost as large as Europe, with a population of 12 million people, in only a few centuries.

Yet scholars insist that the Tiahuanaco culture was the dominant force in the Peruvian highlands during this period. No one has suggested that the Tiahuanaco and the Inca cultures were identical or contemporaneous. Inca legends say that Tiahuanaco was deserted for a long time, and that they did not know who built it, or when, or why. It seems far more likely, then, that Tiahuanaco is extremely ancient—much older, in fact, than early Incas.

In the traditions of the Aymara Indians, the builders were strangers from across the lake. Were these the same people who instructed the natives in the art of government on the island of the sun in Lake Titicaca?

The problem of Tiahuanaco is connected with many ancient ruins in the region—among them the fortress of Sacsahuaman, the cyclopean walls of Ollontay, the great circular stadiums which dot the countryside, parts of Cuzco itself, and the roads.

The network of roads that covered the Inca empire, as efficiently built and organized as that of the Romans, may not, in fact, have been built by the Incas. Inca rulers later took credit for having created all that they took on their path to conquest. Often, however, they inherited from the conquered the things they claimed as their own. It is therefore possible that the Inca actually restored existing roadways, built by the mysterious peoples who had vanished before the Inca empire rose.

The building—or rebuilding—of Tiahuanaco by Wira-Kocha may have been part of a plan to rebuild the civilization that existed before the catastrophe.

On the face of it, the region of Tiahuanaco—bleak, inhospitable, completely barren, and almost treeless—seems an unlikely place to build a city, especially in more recent geological times. Winter is severe, and at such an altitude, the thin air makes any kind of physical exertion a torture to people unused to the altitude. Even the natives resort to a soporific (coca leaves) to alleviate the discomforts of working their lofty homeland.

Before the catastrophe, the situation may have been very different. The distribution of parts of the city suggests that Lake Titicaca was once much larger than it is now, and the presence of large quantities of sea shells on the *altiplano* (high plain) suggest that at one time the whole area was once much nearer to sea level than it is today. The cataclysm that raised this plateau, recent geological surveys indicate, was of fairly recent origin. Perhaps it occurred during the past 6,000 or 8,000 years, when the great disaster wiped out the mammoths, and the polar regions were formed.

The ruins of Tiahuanaco are now some twelve miles from the shores of Lake Titicaca, but the discovery of a dock at the northern edge of the city suggests it was built when the lake was very much larger than it is at present. Furthermore, recent underwater surveys of the lake bottom have reported the presence of walls.

In addition, a wide moat or ditch completely encircled the city, a construction common to many ancient cities, particularly those which appear to have been well planned. This applied to Plato's description of the capital of Atlantis (Crete) and to the city of Jericho.

Tiahuanaco once covered a huge area. Today that area comprises sand-covered mounds and three sections of ruins, known as the Akapana, or Hill of Sacrifices; the Kalasasaya, or Temple of the Sun; and the Tuncu-Puncu, or Place of Ten Doors. The names for these groups of ruins may be hopelessly inaccurate descriptions of the original purposes of the buildings. In addition, the remains of walls, blocks, and statues lie scattered over the area.

The Akapana, the most imposing of the ruins, is a huge truncated pyramidal hill, 167 feet high, and 496 by 650 feet at the base. Each side is almost mathematically in line with the four cardinal points of the compass. Originally the surface was faced with huge smooth stone blocks, but most of these have been hauled away during the construction of the nearby railway or by villagers and farmers. An enormous stone stairway once led to the summit of the monumental hill, but now only a few of the massive steps remain.

Most of the flattened top of the Akapana is occupied by a huge artificial lake with the most scientifically designed and beautifully cut overflow conduits. The original purpose of the Akapana is unknown.

A thousand feet north of the Akapana is the Temple of the Sun, resting on a huge rectangular platform 10 feet above the level of the plain and 500 by 400 feet square. Leading to the terrace, originally covered with stone tiles, is a stairway flanked by huge stone columns. Each step is a massive stone slab 10 feet wide and 20 feet long, weighing between forty and fifty tons.

Along the four sides of the platform are stone columns 15 to 20 feet high and 16 to 20 feet apart. Although many of the columns are worn and eroded, notches and cut mortises are still visible, indicating that the columns were once capped with stone lintels.

Fragments of stone tables, images, statues, and broken pottery have been found in this area, but the most remarkable part of the structure is still almost intact: the famous Gateway of the Sun. Carved from a single piece of exceedingly fine and hard andesite, 13 feet 5 inches long, 7 feet 2 inches tall, and 2 feet thick, with a center doorway 4 feet 6 inches by 2 feet 9 inches, it is the largest single piece of stonecutting in the world.

The western surface of the upper portion has a low bas-relief of geometrical design and four square niches; on the opposite side, the surface is covered with sculptured symbolic figures in low relief. The figures are arranged in a series of squares surrounding a larger central figure representing a sun god (the Weeping God of Tiahuanaco), which holds what appears to be a ceremonial staff in each hand.

Flanking this representation are the forty-eight squares, twenty-four on each side, in three rows of eight. All face the central figure and each carries a small staff. The upper and lower rows comprise identical semihuman figures with wings and crowns; and the central row figures are the same except that the heads are those of condors.

What these figures represent is still a mystery. Professor Arthur Posnansky, who has devoted fifty years to the study of Tiahuanaco, points out that the Kalasasaya was a portion of an astronomical arrangement, and suggests that the figures are calendrical, representing the different months.

Hans Bellamy's book *The Calendars of Tiahuanaco* contains evidence that some aspects of the city's decoration show a year shorter than the present one, and the existence of two moons, but this, like so many other aspects of the city, is the subject of much controversy.

A considerable distance away, the Tuncu-Puncu—the Place of Ten Doors—the largest and most impressive of all the ruins, appears to be a mound of stone 50 feet high and 200 feet square. Actually, the rubble consists of foundation stones, enormous slabs of the fallen walls, stone blocks, columns, cornices, and lintels, all partly covered with debris and drifting sand.

Although its original form is now virtually impossible to visualize, it must have been a huge and imposing building when it was built. It may have been a palace, a court of justice, a public hall, or a temple.

Some of the wall blocks are colossal: many are 36 feet long by 7 feet high, weighing 175 to 200 tons. Bordering some of the largest slabs are huge platforms with seatlike forms whose function is unknown cut into them.

Nearly every large slab that appears to have been a wall section bears deep T-shaped grooves with holes drilled into them at the ends of the grooves. Where two such stones are side by side, these grooves are aligned; apparently the grooves were designed to hold metal staples or keys to bind the blocks securely. A number of bronze and silver staples have been found both in position and lying loose on the site. From their appearance it is obvious that the molten metal was poured into the grooves and hammered to fit.

Pillars and columns were carved with geometrical precision, some deeply, apparently to hold the ends of great stone roof beams, which have been found with their ends cut to fit exactly into the holes. In no case is there a deviation of more than one fiftieth of an inch in the stonecutting, which in a structure of such huge dimensions is fantastically accurate. The flat surfaces are exactly true, almost as if the stone had been planed flat.

Stapling of stone has also been found in some of the ruins of Mesopotamia, and the high degree of accuracy of the cutting reminds us of Egypt's pyramids.

The design and execution of Tiahuanaco's foundations are far ahead of any European city up to recent times. There is a complete system of underground drainage, water supply, and sewage. Throughout the area of the city, perfectly cut stone conduits and pipes are carefully and accurately graded to ensure the flow of the water, and there are sluice gates to control the water in the encircling canal or moat. (The environs of Knossos in Crete, too, have such accurately cut stone drainage and water-supply systems.)

The greatest mystery about Tiahuanaco—common to Stonehenge, the "passage graves," and the pyramids—is the huge scale of the work and the methods by which it was accomplished. *How were these huge weights of stone transported, cut, and erected?* At Tiahuanaco, this work was undertaken under even more difficult conditions than in Western Europe or Egypt.

Many huge blocks in the rough, squared out for transport, have been found in quarries miles distant from Tiahuanaco, some on islands within Lake Titicaca and others on the opposite shore of the lake. The presence of such blocks, roughed out but never moved, suggests that *Tiahuanaco may have been under construction when the catastrophe overtook earth.*

This would mean that construction began at least 6,000 to 7,000 years ago, perhaps earlier, when the civilization in the north was beginning to open up the American continent. Alternatively, the construction may have begun after the catastrophe, and may never have been completed because of the disintegration of what was left of civilization.

If Tiahuanaco had been built by the "heave-ho" technique, using the crudest methods and tools, there is no reason why it should have remained unfinished—indeed, construction of this pattern should have persisted right up to the time the Spaniards conquered the ancient kingdoms. But neither the Spaniards nor any other Europeans saw any megalithic stonework under construction on this scale. All building work the Spaniards observed was small-scale, virtually brick-sized. This small-scale work appears to have imitated the techniques of large-scale stone construction, but not the size.

The Inca did not build Tiahuanaco: they simply did not possess the means to handle such huge masses of stone.

Tiahuanaco's age has always been a puzzle. It appears, for reasons already stated, that the traditional date of around A.D. 1000 is far too recent. The presence of a completely fossilized human skull found among the ruins of the foundations also supports a greater age.

In 1926, 1930, and 1940, astronomers, mathematicians, and civil engineers tried to date the city by measuring the astronomical devices of Tiahuanaco and determining the amount the earth's axis had altered since they were erected. The results obtained were astonishing: Professor Posnansky determined that the second, or middle, phase dated back 13,000 years. Dr. Rudolph Muller, using two different formulas, showed by one method 9,300 years, and by the other method, 14,600 years. Recent estimates indicate the city was built even longer ago—which would make the city an existing settlement *prior to the Flood.*

These calendric calculations assume, however, that the position of the earth has always been constant relative to the visible stars, taking into account only the precession of the equinoxes. If the earth had altered its position due to some *cosmic* event, this, coupled with earth movements which may have shifted the city and its monuments without actually destroying them, may have had the effect of making the city appear much older than it actually is.

In any case Tiahuanaco does belong to the mainstream of megalithic building, which has never been practiced in historic times, and stems from a period before the rise of known civilizations.

The "fortress" at Sacsahuaman stands on an artificially leveled mountaintop above the ancient Inca capital of Cuzco. The structure, covering several hundred acres, consists of three outer lines of zigzag walls surrounding a raised paved area which contains a huge circular stone arena. The ruins also include a complete water-supply system, huge water-storage cisterns, and a maze of underground passages and chambers. Sacsahuaman could have held the entire population of the city it overlooked.

The three rows of walls are built of many angled blocks fitted together without any binding agent, like the pieces of a jigsaw, so accurately that a razor blade cannot be inserted between them. The angled blocks have as many as thirty-six sides, though ten- and twelve-sided blocks are more common. The smallest blocks weigh 50 tons; many weigh 100 tons; others 150 to 200 tons. The largest among them actually weigh nearly 300 tons.

The blocks, carved from an exceedingly hard stone, andesite, were transported to the site from quarries many miles away, one of them on the other side of the river.

From the quarries, they were brought to the top of the mountain, and carefully fitted into place. Judging from their shapes, in some cases they would have had to be lifted, shaped, and then inserted between adjoining stones.

The hypothesis that thousands of men, with the crudest tools, rollers, and ropes, hauled and dragged these stones for years, makes as little sense as the story of the building of the Great Pyramid by the "heave-ho" method.

The upper courses of the walls of Sacsahuaman are layers of small blocks executed in typical Inca style. The water-supply system was kept in working order, and supplies and weapons were stored within the precincts of the "fortress." (The Inca used Sacsahuaman as a fortress and place of refuge for the citizens of Cuzco, but that does not mean that they built it, or that it was originally built as a fortress.) The Inca repaired and maintained the walls in their own style, different from the cyclopean masonry.

At Machu Picchu, the famous citadel perched on a high mountaintop, the same cyclopean stonework appears, with smaller, regular Inca blocks superimposed on them.

If, as the traditionalists maintain, Sacsahuaman was built in its entirety in the fifteenth century, only a few years before the Spaniards arrived, then why could no one remember *when*, and more important, *how*, it was built? Surely memories could not have been so short. The Spaniards were unable to ascertain from the Inca how this staggering feat was achieved.

Legends say that it was built by the gods in a very short time. Parts of Cuzco itself are built in the cyclopean manner, and literally thousands of the gigantic worked stones can be seen in the region of the city. The superstructure of the Temple of the Sun in Cuzco no longer exists, but its floor forms the base of the city's cathedral. This stonework is on the same scale, and has the same hairline accuracy, as that of Sacsahuaman and Tiahuanaco.

These megalithic structures, like those in Europe and elsewhere, must have been built with the aid of highly advanced technological methods of which we are unaware. At all of these sites, little or no evidence of stoneworking equipment exists. In the entire Maya region there is not a trace of a single stone tool which might have been used to carve intricate stone ornamentation.

Amazonian Indian tribes, living in the deep forests close to the borders of former Inca territory, preserve legends of great cities lit day and night by stars, and of a time when men could fly and priests made stones light. The legends, extant when the Spaniards first arrived in the New World, obviously owe nothing to modern inventions.

Many anthropologists believe that the now primitive tribes are the descendants of highly civilized peoples who have, during the course of many centuries, been compelled to lead a more simple existence. The distant ancestors of these primitives may have occupied the fabled cities of the remote past, most traces of which have disappeared.

Whatever their origin, these primitives are not true aborigines. When not exposed to excessive sunlight, they have the pale skin of Mongol people as well as small hands and feet. Many words in their languages seem closely related to the ancient Inca language.

Other peculiarities in the Peru/Bolivia region make very little sense in relation to stone- or early metal-using cultures in a developing state:

• Many of the textiles of the Paracas region, universally recognized as fine examples of textile art, contain as many as two hundred and fifty to the inch, with as many as thirty colors woven into their designs. Modern machine-made fabrics have approximately sixty threads to the inch.

• The lines at Nazca, like Stonehenge and the pyramids, have been associated with spacemen and flying saucers. Situated within a desolate valley, the lines were made by removing the dark surface crust to reveal the yellow sandy subsurface. Varying in length from half a kilometer to more than eight kilometers, they radiate in almost every direction; some are in parallel series, many intersect each other.

There are also solid rectangular, trapezoidal, and triangular forms, one of the largest of which is 1,700 meters long with a mean width of 50 meters. Other lines form spirals and zigzags, and representations of curious birds, spiders, and monsters, some of which seem to have been repeated on Nazca pottery. Some scientists suggest that the lines represent astronomical sighting lines, but this fails to explain the animal shapes and the spirals and zigzags.

Perhaps the most extraordinary aspect of the lines is that *they were detected only when aircraft flew over them.* From the ground the patterns cannot be discerned. How were these patterns and lines drawn? Without scale plans, accurate measuring and surveying devices, and some means to see the completed whole, it is difficult to say.

Von Däniken suggests that the lines represent an airfield, which he presumes was built for, and under the instructions of, visitors from space. This seems logical, since the pattern can be detected only from the air. However, there are many legends of terrestrial flight that owe nothing to visitors from space.

There is a similarity between the lines at Nazca and a navigational device used by Polynesian peoples, which seems to be of great antiquity. The device consists of a framework upon which a series of crisscrossing cords are stretched, apparently at random. In fact, by aligning any given cord with an astronomical sighting point, the device can be used very efficiently to navigate. The Polynesians have used it to navigate from one tiny island to another over trackless miles of the Pacific Ocean.

A string device is used by the Peruvians for another purpose altogether. The *quipu* is a knotted-string device for calculating and recording.

The lines of Nazca may actually be an aerial navigating device, by means of which a flying machine could be oriented along a selected line toward a certain destination.

This thesis does not explain the animal and bird figures; neither do the astronomical or calendric explanations. Perhaps the figures were recognition devices, or perhaps there is no immediate answer to this particular problem. We are not able to examine the psychology or the reasoning of its builders; all we do know is that they must have had a high degree of mathematical knowledge, coupled with accurate surveying and measuring techniques.

• In the Museum of the American Indian in New York there are tiny particles of gold on display. They appear to be natural grains —until they are viewed with a magnifying glass. These "particles" are perfectly wrought beads, some smaller than the head of a straight pin, elaborately engraved, some composed of almost invisible pieces welded together. All are pierced.

The gold beads were found in the region of the Manabis of Ecuador, of whom we have no traces except a great many cut-stone pieces that resemble ancient Roman chairs. These "chairs," some of which weigh up to half a ton, have been found scattered all over the countryside in no set pattern. Their use and function is unknown.

The Manabis left no traces of any pyramids, temples, or other buildings. Whether they undertook the microscopic goldwork is not known. Perhaps this work was left, or dropped, by other people passing through the area.

Of the people who executed this remarkable work, the American archeologist Hyatt Verrill has said, "Either they had lenses, or they had eyes that had the power to see microscopic objects and more adept and delicate fingers than any other race of men." The fact that no lenses or fine metalworking tools have been found supports the contention that the gold beads did not originate in this area.

We are reminded of two other things, however: the crystal magnifying lenses and the Tell Asmar goddess image, both found in

Mesopotamia. This unusual Sumerian figurine is gowned, unlike most Sumerian statuettes, which show a flounced skirt. The image's face has oversized, staring eyes and unusually tiny clasped hands. Why should the Sumerians, who were quite capable of realistic representation of human beings, make such an odd figurine? Do the large staring eyes represent magnifying devices of some sort?

In the older levels of the excavations at Catal Huyuk in Turkey, beads of obsidian have been found with holes drilled through them so fine that a modern steel needle cannot be passed through them. There is no trace of the tools which may have been used to drill through this extremely hard material. These beads apparently stem from the same era as the megaliths.

A further peculiarity of all ancient cultures, both Old World and New, lies in the fact that any given culture usually demonstrates remarkable technical expertise in some fields, and none in others. Some cultures show a great deal of knowledge in, say, mathematics, while others excel in medicine or drainage or plant husbandry.

The Maya, for example, the world's greatest mathematicians, could calculate precise dates for hundreds of thousands of years, and had worked out the exact lunation more accurately than we can today. Yet among the Maya ruins there are no traces of finely appointed villas like those of the Cretans, or fine palaces like those of the Aztecs, and the Maya drainage and water-supply systems cannot compare with those of the Cretans, the Aztecs, or the Inca.

The Inca, apart from their highly efficient form of government, were master agriculturalists who developed many new strains of plants. The little-known Mochica civilization produced some of the most realistic pottery portraiture in the world.

The medical and surgical techniques of the Chimu were probably superior to those of any ancient—and some modern—cultures. Ancient Crete, Tiahuanaco, and the cities of the Indus Valley—Harappa and Mohenjo-Daro—have drainage and water-supply systems superior to anything up to modern times.

On the other hand, ancient Americans, for example, never put the wheel to practical use, and ancient Mexicans never thought of using sails on boats. What was known to one was unknown to another.

It is as if certain areas of knowledge were closed to certain cultures—which makes sense if we hypothesize the collapse and rebuilding of a civilization. Among the flood survivors who created or helped to create new societies, there would be experts in some fields but none in others.

One area, for example, may have had a doctor or group of medical men, but no building engineers. They may therefore have developed remarkable medical techniques, while their city-building was a long, slow process of trial and error. Another area may have progressed in the reverse direction, building well-planned cities and superb drainage systems, while developing hit-or-miss medical techniques.

One society may have had as its core a group of mathematicians, in which case their culture may have developed an extraordinary emphasis—perhaps an almost religious emphasis—on mathematics.

This is in fact what we have found.

In connection with ancient civilizations, we find that the older the culture, the more advanced *some* aspects of its technology appear to be. This could only be so if that technology were, at its oldest stage, nearer in time to the civilization that was destroyed. This would appear to be the only logical answer to the riddles of the ancient maps and the megalithic and microscopic engineering feats.

If our civilization were to be destroyed in a nuclear holocaust, the survivors of such a war, five thousand years hence, would be mystified by the same peculiarities and legends of the emergent civilizations as we must face today in connection with the civilizations of our past.

The Chimu culture of the Peruvian coastal region illustrates many of these peculiarities. We cannot identify the racial type of the Chimu with any certainty; their portrait jars often show people with Caucasoid features, and remains of Caucasoid people have been found buried in ancient graves. Their capital city, Chan-Chan, was well planned and laid out, with efficient drainage and water-supply systems. The people were well dressed and well cared for, but to judge from the vast number of pottery vessels showing every aspect of Chimu culture, the thing they excelled in was medicine, in which field they appear to have been more advanced than either the Egyptians or the Greeks.

Their surgeons appear to have amputated limbs, trepanned skulls, removed injured or diseased internal organs, and filled and crowned teeth. Judging from trepanned skulls which have been found, this operation (in some cases, several operations), was highly successful, as the bone had overgrown the gold or silver plates covering the opening.

Pottery vessels also show people with artificial arms, hands, and legs. It is difficult to see how the Chimu could have attempted such operations without knowledge of antiseptic and anesthetic techniques. Because the Chimu left no written records, they have not received the same credit for their abilities as the more literate (to us) Greeks and Egyptians.

The Chimu may very well have descended from the survivors of a highly civilized people, including a core of highly skilled physicians.

Every civilization of antiquity had an abrupt beginning. Every one appears to have arrived in its particular area from elsewhere. It could be said that *the "culture heroes"* (also known as "culture bearers" or "Flood heroes") *were the first settlers in any particular area, and thus the founders of the civilization.* Like Poseidon of Crete, Gilgamesh of the Sumerians, Osiris of Egypt, and Kukulkan of the Maya, these culture heroes—who appear in the legends of *every* ancient civilization—were later deified by generations of their descendants. The common ancient concept of a divinely descended ruling caste results from this fact.

In the Mediterranean, the Middle East, and Europe, the mythos of the culture bearers has been overlaid by millennia of literate tradition, as well as by superstition and religion. The intermingling of different civilizations and their gods has somewhat clouded the original myths. In spite of these factors, the central figures do show through fairly clearly, and all of these civilizations owe a great debt to the Flood survivors.

In the Americas, the situation is somewhat different. The three great civilizations of which we are aware—the Maya of Yucatán, the Toltec/Aztec of Mexico, and the pre-Inca and Inca of the Peru/Bolivian coastal region—were, from the very earliest times, totally isolated from each other. Still, *all* of those civilizations have two legends in common: the Flood and the culture bearers. To the Mexicans, he was Quetzalcoatl; to the Maya, Kukulkan; and to the Inca, Kon-Ti-ki-Wira-Kocha.

From Mexico to Chile, literally thousands of ruined towns and cities, buried under dense jungle or desert sands, have never been explored. But the legend persists in every tribe, and in the annals of every civilized race in the Americas, of the culture bearers.

They were always described as bearded white men.

In both Old and New World mythologies, the concept of a wise man is an old white man with a long, flowing beard.

Because of the similarities between ancient Old and New World cultures, some authorities believe there must have been a common "mother culture." Usually the Atlantis theory is suggested as evidence. It seems now that the similarities exist because of the culture bearers, but the link is more ancient than has been thought.

There seem to be grounds for suspecting that the antediluvian civilization was widespread in South America, and was almost completely destroyed by the catastrophic events of the Flood epoch. Those who survived reverted to barbarism, probably within a fairly short period—perhaps several centuries.

The almost complete return to savagery was not broken until the culture bearers either established new centers of civilization as in Crete or attempted to resurrect civilization among the scattered, now barbaric survivors.

Several theories have attempted to explain the rise of the American civilizations.

Rome's rule of Britain collapsed in the fifth century A.D. By the eighth century the Romans had been almost totally forgotten; their towns and cities were obliterated, and the inhabitants of these islands were reduced to barbarism, torn by intertribal strife and assaults from the Continent. From orderly civilization to almost complete barbarism in only three hundred years—and this in an area of the world where the light of civilization was only dimmed, not extinguished.

One theory, popular in the nineteenth century and still supported in some quarters, is that the Maya civilization was created by an emigrant tribe of Israelites. This theory is still strongly supported today by the Mormon Church, which cites the Book of Mormon as its authority.

Both the Book of Mormon and the Maya *Popol Vuh* describe the Flood and widespread destruction. The Book of Mormon mentions that a great deal of cement was used in the construction of Maya cities; archeologists investigating Maya sites confirm this. The Book of Mormon is alleged to have been transcribed from forty-eight metal plates; a Maya legend relates that the entire history of the Maya was written down on forty-eight golden plates fastened together to form a book, which was hidden by the priests when the Spaniards arrived. This fabulous golden book has never been found.

Unfortunately, some things in the Book of Mormon do not seem to connect with Central America during the past 3,000 or 4,000 years, when this migration of Hebrews was supposed to have taken place. For example, it mentions certain animals in the "new land" that do not live in Central America, and does not mention things that do exist in Central America.

Although most authorities outside the Mormon faith dismiss their claims, it does serve to illustrate the widely held theory of a "transplantation" of Old World cultures to the Americas. Perhaps both Hebrews and Americans *were* descended from a common source, much further back in time.

Another theory contends that the American cultures were the creations of Egyptian colonists or visitors. As evidence, its adherents cite step pyramids and commemorative stelae, common to both Egypt and Yucatán. Should we not therefore expect *more* concrete evidence of Egyptian contacts—writing in hieroglyphs, representation of some of the principal Egyptian gods, or the practice of mummification? (Mummified corpses have been found in the Peruvian coastal area, but they owe nothing to Egyptian preservative techniques; for the past several thousands of years at least, parts of the Peruvian coast have been virtually rainless, and this has preserved the corpses.) Of course, we cannot entirely rule out the possibility of trans-Atlantic contacts between ancient cultures on both sides of the Atlantic, but it would seem that these contacts would have been much more recent.

The suggestion that the Vikings could have been the white culture bearers is based on faulty chronology: the Norsemen visited the northern part of the American Atlantic coast several centuries after the start of the Christian era, and the American pre-Columbian cultures are known to have been in existence in 2000 B.C. As in the case of the Egyptian hypothesis, we should have traces of Nordic gods and perhaps Nordic influence in building. These are completely absent. Furthermore, the Norsemen during this period were neither mathematicians nor builders of cities, while the Maya were both.

We must look elsewhere for the culture bearers of the Americas. One thing is certain: *in all the legends of the Americas, the white culture bearers were bearded and had Caucasian features.* Sometimes they are described in the singular, sometimes in the plural, which suggests that they were not single invincible beings but groups, in many cases, of what appear to be men with extraordinary powers.

There is some degree of convergence between Old and New World legends of the culture bearers, although the details differ.

Ancient Egyptian representations of gods and deified kings, for example, always show them wearing false beards. This attachment of a false beard has always been a sign of divinity throughout the Mediterranean basin, as well as Egypt. In these areas reddish or blond hair was also a sign of godliness. The ruling classes often dyed their hair red with henna as a symbol of divine descent. The mummy of a princess found in Egypt still had hair adhering to the skull, preserved in the dry climate for thousands of years. The hair was auburn in color and wavy, betraying a European origin for at least the most influential of the Egyptian populace.

In the New World, portrait jars of Chimu chiefs from the ruins of the city of Chan-Chan on the Peruvian coast show men of a completely European type, who might have been Greeks or Romans. Mummified corpses of chiefs from the oldest layers of graves in this region bear hair that is auburn or blond, wavy, and fine—the opposite of Mongoloid hair, which is always black, straight, and coarse by comparison.

Legends about the white culture gods are more prominent in those areas where urban cultures existed—Mexico, Yucatán, and the Peruvian coast—than among the natives of North America or the southernmost part of South America.

At the time of the conquest of Peru, the Spaniards noticed that many of the Inca ruling caste were paler of skin and had reddish tints in their hair, as distinct from the native mountain peasants of the Andes, who were generally of distinctly Mongoloid ancestry. It appears that an effort had been made down the centuries to keep the white strain as pure as possible.

When Captain Cook first explored the South Sea islands, which at that time had had little contact with the rest of the world, he observed how fair-skinned the native Polynesians were. Many had reddish hair.

From the north to the south, we have the American legends of the white culture gods:

Mexico: The Aztec historians remember Quetzalcoatl, the great priest-king of Tula or Tollan (not to be confused with the present town of Tula), teacher of the arts, law-giver, virtuous prince, master builder, and merciful judge, who came with his builders, astronomers, mathematicians, and painters.

Quetzalcoatl was wise, kind, and humane, gave maize and other foods, and taught that the divine powers required no sacrifices of either men or animals, but offerings of fruit, bread, or flowers. The Aztec historian Tezozomoc wrote, "Where the bright sun lives [Tula] and where the god of light [Quetzalcoatl] forever rules as long as that orb is in the sky."

Quetzalcoatl, we note, did not work unaided, but was accompanied by others of his kind. In Aztec legend, the Toltecs, or perhaps a people who preceded the Toltecs, were the master builders of ancient Mexico, and it was a matter of pride to be able to trace ancestry to Toltec origins.

The greatest extant memorial of the Toltecs, with origins extending far beyond the Toltecs (who preceded the Aztecs), was the city of

Teotihuacán, near what is now Mexico City. The present complex includes the largest pyramid in the world, exceeding in bulk the Great Pyramid of Egypt. The Pyramid of the Sun, like other American pyramids, was stepped, with a great stairway leading to its flattened summit. Many small structures line an impressive ceremonial way that leads to a smaller Pyramid of the Moon.

> Among Maya pyramids, the stairways were made wider at the top than at the bottom, eliminating perspective. Looking upward, the illusion is created that the stairway is the same width at the top. We noted this use of entasis at Stonehenge.

A large secular town lay beyond the boundaries of the huge ceremonial enclosure, with an estimated population of half a million in its prime. This great city, *crumbling into ruin when the Aztecs were raising their own city of Tenochtitlán,* was surrounded by legends, one of which said that it was the place where the gods gathered to judge the affairs of men.

Although there are certain Toltec elements in the city of Teotihuacán, the city is thought to be far older than the Toltec period, and has an unknown ancestry. Did the original Quetzalcoatl build this city? Is it the Tula of the legends?

No other site in Mexico seems to fit the description. This would make Quetzalcoatl much older than scholars have thought: in ancient Mexican history it is said that Quetzalcoatl and his followers were driven from the city and appeared to have gone to Yucatán, which is reputed to explain the Toltec influence at Chichén Itzá. However, it was common practice in the Americas, as elsewhere, to hand on the style and title of some great person for many generations; at the time of the Spanish conquest, some Indian tribal chiefs were still called Quetzalcoatl. The first Quetzalcoatl may have been from a very remote time indeed.

Yucatán: Maya legends say that the first civilizations were created by the "Old White Fathers," the Saiyam Uinicob, who came in the Great Arrival from the sea.

They were also called the "adjuster-men." Why? What did they adjust? Were they the people who *adjusted the calendar*?

Maya mathematics and calendric computations reached a degree of perfection unmatched anywhere else on earth even to this day. The creators of the Maya civilization may have been the mathematicians and astronomers who came from the east among the Flood survivors, who taught mathematics and reorganized the calendar after the catastrophe. This may eventually have assumed the character of a sacred art, forming the basis of a religion based on mathematics.

There have been stranger reasons for creating a religion. In modern times, cargo cults have sprung up in New Guinea, based on the sightings of aircraft by natives.

There is an affinity between the writings of the Aztec historian Tezozomoc and Maya legends, which is all the more remarkable when we consider that the two cultures never met, and that the Maya civilization had already ceased to exist when the Aztecs became a nation.

According to one Maya legend, "Under the beneficent rule of Kukulkan the nation enjoyed the halcyon days of peace and prosperity. The harvests were abundant and the people turned cheerfully to their daily tasks, to their families and their lords. They forgot the use of arms, even for the chase, and contented themselves with snares and traps."

Another legend reports that their whole civilization was built by a single genius: was this Kukulkan? It seems more likely that it was built by a group of survivors from the catastrophe.

Peru/Bolivia: In the great Inca empire, which covered a vast area of the South American Pacific coastal area and the highlands of the Andean *altiplano*, the legend of the white culture gods was widespread. (Like the Aztec, the Inca thought that the coming of the Spaniards was the return of the white gods from the sea.) In Peru, the culture god was Wira-Kocha, or Kon-Tiki-Wira-Kocha. To this day white men are called *virakochas* by the Andean natives of remote villages.

There is a great mystery about the beginnings of the Inca empire. We have already touched on the difficulties involved in a theory that the rise of the Inca empire preceded the arrival of the Spaniards by only a few centuries. This would mean that the entire Inca system of government for a vast area of difficult territory, the organization of all its peoples and resources, and the institution of a highly complex and efficient administration, would all have been accomplished in the course of a few hundred years. In addition, the Inca introduced a separate language, a sort of Esperanto, called Quechua, to replace dialects used among many different nationalities and tribes. How this could have been done in an area where the rulers themselves were natives is difficult to understand.

The legends of the Inca tell a different story:

"[One] legend speaks of certain white and bearded men, who, advancing from the shores of Lake Titicaca, established an ascendancy over the natives and imparted to them the blessings of civilization. It may remind us of the tradition existing among the Aztecs in respect of Quetzalcoatl, the good deity, who with a similar garb and aspect, came up the great plateau from the east on a like benevolent mission to the natives. The analogy is more remarkable, as there is no trace of any communication with, or even knowledge of, each other to be found in the two nations." (Quoted by William H. Prescott in *Land of the Incas.*)

Bandelier was told the following story by an "old native wizard":

"In very ancient times, the Island of the Sun [in Lake Titicaca] was inhabited by gentlemen [caballeros] similar to the *virakochas* [white men]. Whence the gentlemen came, he knew not. The gentlemen had intercourse with local native women and their children became the Inga-re [Inca] and they [the Inca] drove away the gentlemen."

Betanzos tells of still another legend:

"When I asked the Indians what shape this *virakocha* had when their ancestors had thus seen him, they said that according to the information they possessed, he was a tall man in a white vestment that reached to his feet, and that this vestment had a girdle; and that he carried his hair short with a tonsure on his head in the manner of a priest; and that he walked solemnly, and that he carried in his hand a certain thing which today seems to remind them of a breviary [book] that the priests carry in their hands. When I asked what this person called himself they told me his name was Con Ticci Virakocha Pachayachachic, which in their language means God, Maker of the world."

When "Con Ticci Virakocha's" sculpturing at Tiahuanaco was finished, Betanzos tells us, he is said to have ordered all his original followers but two to go away. He told his departing *virakochas* that they were to observe the Tiahuanaco statues and the names he gave to each kind.

Pointing to the statues, he said, "These should be called so and so and should appear from such and such spring in this or that district and should inhabit it and multiply there; and these others should appear in such and such cave and should be turned so and so and settle there and there; and such as I have pictured them and made them in stone they should appear live from the springs and rivers, caves and mountains in the provinces I have told you. And afterward you should all go in that direction"—he pointed toward the east—"and spread them out, showing them the road each of them is to take."

The legend contines, "Virakocha arrived at a province called Cacha, eighteen leagues from Cuzco, and the Caras Indians were armed and did not recognize Virakocha. They came in a crowd to kill him, and when he saw why they were coming he caused fire to fall from the skies and begin burning a hill near where the Indians stood. When the Indians saw this they threw away their weapons and prostrated themselves before Virakocha, who beat out the fire with a staff."

On this spot the Indians erected a stone statue of Wira-Kocha twelve feet tall. This statue of a bearded man was in existence at the time of the Spanish conquest. Cieza de Leon, a post-conquest chronicler who saw it before it was broken up as a heathen idol by fanatical Spanish clergy, confirms that the figure was of a European-featured man with a beard, and adds that the garments held marks to indicate they appeared to have been fastened by buttons.

The Incas believed that after the creation, Wira-Kocha sent a great flood to punish the first men, but a few were preserved to repopulate the world. When the flood was over, Wira-Kocha appeared suddenly on Titicaca, to help restore mankind and give them light.

The *quipocamayoes* (Inca recorders) described to Sarmiento de Gamboa, author of *History of the Incas,* the appearance of Wira-Ko-cha: "A man of medium height, white and dressed in a white robe, with an alb secured around the waist. He carried a staff and a book in his hands."

Some of the Tiahuanaco statues carry in one hand what can only be a book, which appears to be fastened with clasps. Even as "late" as 2000 B.C., books were unknown in the Old World, and writing was confined to either clay tablets or scrolls. Yet we observe that the Indians of the Americas had a long tradition of paper records in book form.

Oddly enough, in the area of Inca domination, writing was apparently unknown, books were nonexistent, and records were kept by means of the *quipus* (knotted strings). Apparently the Wira-Kochas did possess books. This may be a clue to their great antiquity. Perhaps by the time the Incas came to power, the art of writing had been long forgotten.

Cieza de Leon recorded this story:

"They tell that from the south [of Cuzco] there came and stayed a white man of tall stature, who in his appearance and person showed great authority and veneration; and that as they saw he had great power, turning hills into plains and plains into hills, making fountains in the solid rocks, they recognized such power in him that they called him the creator of all things, Father of the Sun, because, besides this, they said that he made greater things, as he is said to have given men and animals their existence, and finally that wonderful benefits came from his hands.

"In many places they tell how he gave rules to men, how they should live, and that he spoke lovingly to them with much kindness, admonishing them that they should be good to each other and not do harm or injury, but that instead they should love one another and give charity.

"He went off northward while accomplishing these wonders and was not seen again."

Another legend says that after the Wira-Kochas had imparted their wisdom, they went on foot to the seashore at Puerto Viejo on the equator, and vanished into the west, walking on the sea.

Although many Inca legends are clear that the Wira-Kochas left Peru across the Pacific by sea, it is never made clear how they arrived at Lake Titicaca. Some legends said, as we have seen, that they simply appeared. One describes an arrival on a "gold disc" from the sky.

Von Däniken says in *Chariots of the Gods?* that according to legend a woman called Oryana arrived in a golden spaceship from the stars and returned to the stars when the purpose of building the city of Tiahuanaco and producing earthly rulers was accomplished. This Oryana was human in appearance but had hands with *four fingers*. The representation of the Weeping God, the central figure on the Gateway of the Sun at Tiahuanaco, has four-fingered hands.

These legends lend themselves to an interpretation by certain Christian sects that Wira-Kocha was in fact the risen Christ on a mission to the New World, or that these civilizers may have been early Christian monks. Unfortunately, it appears that all this transpired long before the arrival of Christianity or its founder. (The original legends of Wira-Kocha may have been overlaid with a veneer of Christian lore, particularly if those who recounted the legend to the Spanish conquerors had been converted to Christianity.)

Many scholars dismiss the Wira-Kochas as mythical beings, and the white skin and beards as symbolic representations of light and the rays of the sun. But all the legends of both Middle and South America oppose this. There is nothing supernatural in the descriptions of the culture bearers: they do not glow or shine, they are not misty or ghostlike. They could be killed or driven away.

The physical evidence supports the myths: the representations in stone of white bearded men, long before the Spaniards set foot in the Americas, and the skeletons from the Chimu graves, which show distinctly European characteristics.

The legends of the Toltec/Aztec region of Mexico and the Maya of Yucatán, and those of the Peru/Bolivia civilizations that culminated in the Inca empire, share a remarkable unanimity of theme: *all tell of white, bearded gods with remarkable abilities, who taught peace and brotherhood.* Yet never at any time did these widely separated cultures meet, or know of each other's existence.

Aspects of the culture-god theme are common to both the Old World and the New. The term Inca, for example, which has been applied to the whole of the people of the Inca empire, actually refers only to the ruling class. More accurately, the term is Inga-Re—"Children of the Sun." The sun deity of the Egyptians was Ra or Re, and many of the pharaohs incorporated this title in their name, to identify themselves with the sun god.

Both Quetzalcoatl in Mexico and Wira-Kocha of South America are referred to as "bringers of light" and are associated with "cities of light." The Old Testament refers to the New Jerusalem, the "City of Light"—which is also mentioned in the New Testament, especially in the book of Revelation. The eventual return to the "City of Light" can be interpreted as a reference to the restoration of paradise. It can be argued that both the biblical statements and the American legends refer to the "cities of light" that existed in ancient times.

We know that the megalith builders and the pyramid builders of the Old World lived during a period we call prehistoric; it is therefore logical to assume that the culture bearers of the Americas, who also built pyramidal and megalithic structures, were their contemporaries. *The culture gods of the Americas and the culture gods of the Old World stemmed from the same source.*

One of the most outstanding features of all religions and religious literature of great antiquity—Sumerian, Hindu, Judeo-Christian, Persian, and the religions of the Americas—is that *all have as a basis a catastrophe of enormous proportions.*

They share these themes:

Man was created or placed on earth to rule and look after it. In the course of time, mankind multiplied and became too clever and/or too wicked, and rivaled the powers of the gods.

Because of this wickedness or their challenge to the gods' power, the gods decided to destroy mankind, except for a few, who were warned of the coming destruction.

This destruction took the form of a great flood, great earthquakes, winds, and rains of fire. In the wake of this catastrophe, the earth's climate was changed, man's life span decreased drastically, and he became prey to many sicknesses. He no longer had direct communication with the gods.

Other, subsidiary, themes are quite widespread: the widely diffuse theme of the Divine Scapegoat, a sacrifice to appease the anger of the gods, which has to do with a sense of universal guilt; the sacred water ritual; the concept of the devil, banished from heaven to earth; and the eventual return of the gods from heaven to earth.

It is our contention that *all the traditions and legends contained in religious literature are based on events in the past and contained warnings to future generations*. Likewise, some religious rituals grew out of actual events and needs. With the passage of time, these events took on the cast of supernatural events, and real people and the things they did became gods and miracles.

Many myths contained in religious literature and tradition make little sense unless thay are taken within the context of a collapsed civilization and a great catastrophe. But other attempts have been made to explain them.

The most widely disseminated point of view is that most religious writing, from the Indian epics to the Bible, either comprises fables to express certain "truths" or prescribe an ethical code or are the results of overworked imaginations, sometimes brought about by drug-induced hallucinations.

It has been suggested that Judaism, and therefore other religions, is rooted in fertility cults whose members were drug addicts. (See John Allegro's *The Sacred Mushroom and the Cross*.) A variation of this thesis suggests that the prophets, including the figure of Christ, suffered from hallucinations brought on either by fasting or by deliberate or accidental drug-taking.

However, much of the Old Testament describes events and places, dress, law, religious institutions, and taboos with a great deal of historical accuracy, *verified by archeological discoveries in recent years*.

The Exodus story of the plagues of Egypt, for example, has often been interpreted as symbolic of the Hebrews' escape from bondage, or as propaganda to support their view of the superiority of their God over those of other nations. In view of the discovery of the Thera eruption in 1500 B.C., this Exodus episode has been reevaluated. Hebrew scrolls which were not included in the Old Testament show that three quarters of the Hebrews died in the aftermath of the eruption. This episode has moved out of the realm of fable into that of real events. Only the *causes* have been ascribed to a supernatural source.

How many other curious biblical stories also describe real events? In one episode, for example, Hebrew priests were said to be consumed by a strange fire. Was this caused by a seepage of crude oil, ignited by the fire the priests made on their altar?

Could fire have been "conjured up from heaven" by the use of a lens concentrating the sun's rays?

Many Old Testament events attributed to supernatural causes were distorted versions of real events, some of which were undoubtedly designed to maintain priestly power and perpetuate fear of God.

UFO phenomena have recently been the subject of serious study. The Old Testament story of the prophet Ezekiel has frequently been cited as a close-range, prolonged sighting of an extraterrestrial vehicle. That the occupants of the strange craft are not described as either gods or angels, but as metallic creatures with the look of human beings, suggests that the prophet was not quite sure what or who they actually were.

His account of wheels that moved of their own volition, making a sound like rushing waters, could have described a machine. At a time in which few metals were known and all conveyances were powered by either humans or animals, something like a powered machine would appear miraculous. Ezekiel would also encounter great difficulty in describing it.

Many of these stories and "visions" are close parallels to basic myths and religious literature from other ancient major religions. For example, the Ezekiel episode seems to be connected with the idea of flight. He says he was lifted to a great height and flown to a "frame of a city" in the mountains. The description of this "city frame" might refer to a spaceship of considerable size, perhaps under repair. There is also a reference to a chamber Ezekiel must not enter on pain of death; even the priests have to wear special clothing. Perhaps this was the reactor room, with attendant radiation hazards.

In one Sumerian legend, a man was taken up to heaven in the claws of a huge eagle, where he became very heavy as he ascended. Could this be the "G-effect"? From a great height, he noted that the ground looked like porridge—a fairly accurate description.

Flying machines, called *vinamas*, appear often in Hindu Vedic lore. They appear to have been widely used as transport by the gods.

The concept of flying, then, is common to legends of every major ancient culture, and is not the product of the "hallucinations" of any particular sect. *Some of the descriptions so closely resemble modern experience that they would seem to be based on a core of reality.*

Apart from straightforward history and religious rhetoric, much of the Christian Bible has to do with the theme of catastrophism.

The Exodus episode, which we have discussed at length, details certain catastrophic events connected with a vast eruption in 1500 B.C. The events of the Flood and the Fall are not so clearly delineated.

Is the Exodus episode more detailed because the chroniclers were nearer in time to this event? Is the story of the Flood less clear because the events were as far away in time from the chroniclers as the events of the Thera eruption are from us? This added temporal distance makes them no less *true;* what has placed them in the realm of myth is their lack of clarity due to their remoteness in time.

It also appears that biblical accounts of the catastrophe are rather disjointed, as if different chroniclers had remembered odd incidents. The episode of Sodom and Gomorrah, for example, would appear to be connected with the catastrophe, as would many references in Isaiah.

The culmination of the catastrophe motif in the book of Revelation is perhaps only partly a vision of the future; some parts may refer to the past. The war in heaven, the banishment of Satan, the great flying metallic creatures—was the war in heaven a battle fought high in the stratosphere, or on the fringes of space? Were these creatures aircraft armed with weapons?

So many events in the Old Testament can be related to actual events. Let us look at the more apparently mythical episodes as if they, too, might refer to actual events.

Let us assume that humanity at this time was a colony of an extraterrestrial power. If a Superior Community had discovered a planet suitable for colonization—in this case, earth—it would perhaps have prepared it for colonization, using what we now call "planetary engineering" to create a more suitable environment. This would account for the statement in Genesis that the world was made for man and that he was given dominion over it.

Perhaps there was a revolt ("disobedience") against control by the Superior Community—or perhaps the colonists discovered some weapon or planned to embark on a campaign of conquest once they had become numerous and strong enough. Such assumptions would account for the legends of the gods' fearing man because of his attempted rivalry. Legends of communication between man and God in the antediluvian world could refer to channels of communication, severed in postcatastrophe times.

The theme of Lucifer, once an angel, who rebelled against God and was cast out of heaven, is often thought to symbolize the human conflict between good and evil. Let us look at it from another perspective:

Lucifer (mankind) rebelled; by destroying his civilization, the rebellion would be quashed. The ability of the human race to travel in space (heaven) would be eliminated if the necessary technology were lost, and thousands of years would pass before the secrets of space flight were rediscovered.

The Hebrews regarded themselves as the Chosen People. Why?

As history, the Old Testament is the most complete overall account of these events. Other religions carry sometimes very detailed descriptions of certain events, but lack this overall view.

Perhaps one group was entrusted with the task of recording the circumstances leading to the catastrophe and the effects it would have on the future of the human race. *The Hebrews may have been "chosen" to transmit the history of the catastrophe.*

Certain stories recounted in the Old Testament make little sense if we assume they arose out of a local religion with a primitive god who differed little from a host of other vengeful deities. For example, the kind of knowledge implied in the statements that God would "move the earth out of her place," and "the moon and the earth, they hang upon nothing" would not have been the province of chroniclers drafting the Old Testament in its present form.

The sacred water ritual is common to many religions, including Judeo-Christian, Indian, and Inca. Baptism and the washing away of sins in the Inca religion were so startlingly similar to Christian practices that a Spanish priest was heard to observe that the Incas were not heathens at all, a remark which cost him his life for heresy.

Washing with water to cleanse from sin is such an ancient practice that it is difficult to trace it to any source. Although by historical times it had become a ritual, it must at some time in the past have had a practical function.

References in the *Mahabharata* indicate that the only way for people to escape sickness in the aftermath of the weapons was to strip off their clothing and wash themselves in the river. The running water of rivers also figures prominently in religions of historical times: the Jordan, used by Hebrews, the Ganges in India, and streams of water conduits by the Inca.

The cleansing of sin by water was introduced to the Inca by the god Wira-Kocha; the "sacred spring" and its temple exist on the Island of the Sun in Lake Titicaca. Particularly if there had been an amount of residual radioactivity in many areas, Wira-Kocha would have warned survivors that to avoid contamination they should immerse themselves in running water.

The concept of the dying and resurrected god, the Divine Scapegoat, is common to the Judeo-Christian, the Greek, the Egyptian, the Inca, and many other ancient American religions. Even the sacrificial rituals are similar: for example, in the Greek cult of Dionysus, the blood and flesh of the bull represent the blood and flesh of the slain and resurrected god; in the Christian faith, the bread and wine represent the body and blood of the crucified and risen Christ.

In the Inca religion, the Supreme God Camac, represented by the sun, with his son, Inti, are killed (at sunset) and resurrected (at dawn). Bold Spanish clergy drew attention to the similarity of this religious doctrine to the death and resurrection of Christ.

In fact, most death-and-resurrection religion seems to have a strong connection with solar worship. Christ said, "I am the light of the world." There seems to be a connection between this Christian mythos and the widespread concept of the "Sacred Children of the Sun."

Why should a religious pattern develop involving worship of the sun coupled with a fear that it may not return? *The feared death of the sun and the dying and resurrected god are intimately connected.* Although we have no evidence of religious forms from the "prehistoric" period, it seems such notions may not have existed prior to the catastrophe—*because there would have been no reason for them.*

The dying-and-resurrected-god concept may also relate to the concept of universal guilt and a need for atonement for whatever man did to bring about the catastrophe.

The movement of the earth away from the sun would have altered the climate of the planet, suddenly and drastically in some areas. The attendant disturbances may have completely obscured the sun, perhaps for a long time in some latitudes. This would account for legends that the sun disappeared for a long time, and that when it returned, it was farther away.

It would be quite natural for the survivors and their descendants to fear a repetition. For this reason the daily sunrise was welcomed with great relief; it also explains the concern that at the end of a given period of time there would be a recurrence of the catastrophe. The Aztecs and Incas dreaded the ending of the fifty-two-year cycle, and sacrificed to ensure that the disaster did not recur.

The changes in climatic conditions may also explain the origin of fertility cults. The sudden, intense cold in formerly warm lands, coupled with the death of vegetation—and human population—that such changes would bring, would have seemed totally unpredictable. The people most closely involved were not to know that these conditions would change for the better. The appearance of spring, heralding a new cycle of seasonal change, must therefore have been cause for great thanksgiving. Both the continued reappearance of the sun and the coming of spring after the winter "death" became times of rejoicing and sacrifice to ensure that spring and harvest would continue and the sun would continue to shine.

A major theme of all religions is that before the Flood the world was ruled by gods. These long-lived ones before the Fall would have seemed godlike to those who came later, and lived shorter lives. The pre-Flood people, free of sickness and disease, who commanded powers undreamed of by later ages, must have seemed to be gods, although in reality they were no such thing.

Together with a multiplicity of gods who often display surprisingly human behavior, there is also the concept of an *invisible god* who permeates the whole of existence, and is responsible for the existence of all living things. It may be that the concept of a "universal force" derived from those who lived before the Flood; this memory, together with the memory of the godlike men, may have been incorporated into later, postcatastrophe religions.

Recent advances in science, particularly in the field of nuclear and subatomic physics, suggest that some reversal of the materialist view of the universe may be in order. The universe, it is now thought, may not be intelligible primarily as "matter," but rather as particle states and energy forms. In short, there may be more to life than chemical interactions.

The scientific concept of a fundamental particle of mental energy (mindon), which activates chemical systems so that they become life systems, is extraordinarily like certain basic religious concepts. God is frequently represented as the life-giver, breathing the spirit of life into inanimate objects, or the spirit, or universal mind, which has given all creatures their life.

Are not these two expressions, the one religious, the other scientific, the same thing? Much religious thought and writing is being borne out by discoveries in every field of science.

Religion, far from being at odds with science, may itself be a vague and distorted memory of things once known, things we are now beginning to rediscover through science.

A CONCLUDING WORD

If it were true that mankind took a fatal step some thousands of years ago that led to his virtually complete destruction, toppling a great civilization, we are left with a sobering thought: if we, who now have the capability to unleash unlimited destruction, were foolish enough to do so, we could plunge humanity into a new barbarism that could last ten thousand years.

It may have happened before. Let us hope it does not happen again.

INDEX